U0173827

[美国]道恩·E.霍尔姆斯 著　李德俊 洪艳青 译

牛津通识读本·

大数据

Big Data

A Very Short Introduction

译林出版社

图书在版编目（CIP）数据

大数据 ／（美）道恩·E.霍尔姆斯（Dawn E. Holmes）著；
李德俊，洪艳青译. —南京：译林出版社，2020.10
（牛津通识读本）
书名原文：Big Data: A Very Short Introduction
ISBN 978-7-5447-8343-9

I.①大… II.①道… ②李… ③洪… III.①数据处理 IV.①TP274

中国版本图书馆 CIP 数据核字（2020）第 129444 号

著作权合同登记号　图字：10-2018-429 号

大数据 [美国] 道恩·E.霍尔姆斯 ／著　李德俊　洪艳青 ／译

责任编辑　　陈　锐
装帧设计　　景秋萍
校　　对　　戴小娥
责任印制　　董　虎

原文出版　　Oxford University Press, 2017
出版发行　　译林出版社
地　　址　　南京市湖南路 1 号 A 楼
邮　　箱　　yilin@yilin.com
网　　址　　www.yilin.com
市场热线　　025-86633278
排　　版　　南京展望文化发展有限公司
印　　刷　　江苏凤凰通达印刷有限公司
开　　本　　635 毫米 ×889 毫米　1/16
印　　张　　16.75
插　　页　　4
版　　次　　2020 年 10 月第 1 版
印　　次　　2020 年 10 月第 1 次印刷
书　　号　　ISBN 978-7-5447-8343-9
定　　价　　39.00 元

序　言

王崇骏

四方上下曰宇，往来古今曰宙！

从137亿年前宇宙大爆炸，到46亿年前地球诞生，再到38亿年前地球上开始有生命，直至约6 000年前文字的出现开启了人类文明，这是一个多么漫长的过程！与其他物种大多通过遗传进化不同，人类在进化过程中发展和演化出了一种非遗传性继承：通过独一无二且日益发达的文化媒介（语言、文字以及有意识地利用外在物和工具的特质）将知识留给后代。这种文化传承使得人类可以快速进化，并最终成为这个星球的"统治者"。

在漫长的人类文明发展史中，从源自动物性的"数觉"到为了"征服"自然界，人类开始对"数"产生了需求，并且随着这种需求的逐步膨胀，一系列的工具、算法、设备被不断发明和创造，比如早期的计时工具、计数工具，以及诸如算筹、算盘等算术工具。随着计算理论的丰富，以及机械工艺的进步，人们开始思考如何用更为精巧的设备进行计算，如纳皮尔筹、机械尺、机械计算机等；而图灵机理论的发明、冯·诺伊曼体系结构的提出，以及1946年ENIAC的发明，则让数字计算时代就此来临。

在后ENIAC时代，计算机从最开始的军用走向了民用，其功能不断发展和丰富，从最开始的数值计算走向网络通信、计算感知，并逐渐应用于生活娱乐和企业管理。与此同时，也因为计算机在各个领域的渗透和深入应用，计算机从最原始的计算工具变为一个研究对象，伴随着各类计算理论的发展，计算机科学与技术、工程和应用也得到持续的发展和推进。在这个发展过程中，传统的"数"的内涵，也从最开始的"数值"不断拓展，数据开始以不同的类型、模态、视图样式出现，并服务于人们的需求，传统意义的"数值"则成为一种数据类型。

信息技术，尤其是互联网技术的迅猛发展，烟囱式软件开发模式、云计算在不同领域的不断渗透，以及人们日益多样化和碎片化的行为方式，或许还有其他更多的原因，让人们在数据层面不得不面对"大数据"这样的难题，即难以在期望的时间内利用常规工具进行有效处理并获得期望的价值。在当下的大数据时代，传统意义的"数据"成为一种可以交易的"资产"，一种具有战略资源意义的"石油"，一种提高竞争力的"资本"，一种用于科学研究的"第四范式"。出于对大数据价值期望的共同追求，社会各界都对大数据产生了极大的兴趣、热情和期盼，使得"大数据"这个概念从其诞生之日起，就得到了"政、产、学、研、商、用"的一致认同，并引起了包括哲学家、科学家、技术研究者和工程研发人员等的普遍关注。

牛津通识读本《大数据》一书，当然也是对这一社会关注点的回应。道恩·E.霍尔姆斯教授是贝叶斯网络、机器学习和数据挖掘方面的专家，她用深入浅出的文字扼要介绍了什么是数据、什么是大数据以及它有什么意义，进而对大数据应用所涉及的存储和分析技术进行了简明扼要的综述，并在此基础上分析和研判了大数据在医学、电子商务、安全、生活等各个领域的应用。这样一本概述性质的通俗读物，非常有助于普通读者更多地了解大数据，学习大数据，应用大数据，进而培养一种大数据思维，以适应未来的大数据时代。

对于本书所涉及的内容，笔者还想稍做一些补充，即看待大数据的三个视角和四个维度，以帮助读者更加全面地认识大数据。首先谈谈三个视角。

一是计算视角：从计算视角来看，大数据是一个难以获取、难以组织与管理、难以处理和分析的技术难题（以及因此而引发的各类思维层难题），也正是因为这样的难题驱动，加之人们对大数据在优政、兴业、科研、惠民等不同领域的价值期望，促使相关科研人员进行技术攻关和发明创造，进而推进了相关理论和技术的发展。

二是科研视角：从科学研究的角度来看，大数据成为继实验、理论、模拟之后用于科学研究的"第四范式"（此处的"范式"指的是从事某一科学的科学家群体所共同遵从的世界观和行为方式）。

三是商业视角：从商业应用的角度来看，大数据能够带来利润。一般而言，只要找到一个合适的应用场景，并为这个应用场景找到一个合适的解决方案，知道数据的来源并且能够获取，而且有技术支撑（研发能力），更重要的是能够找到融资支持（经过可行性分析、盈亏平衡分析等之后），就有可能最终成功应用并获得收益。这在彰显大数据商业价值的同时，也会促使同行去挖掘更多的大数据价值。

更进一步说，大数据的价值实现涉及数据、技术与应用的协同，具有典型的多学科交叉与跨界整合特征，因此就总体而言，实现大数据价值至少涉及如下四个维度（层面）。

1. 算法（Algorithm）：大数据价值的实现路径涵盖了数据采集与汇聚、数据存储与管理、数据处理与分析、应用系统开发与运维，每一个环节都需要依赖不同的算法进行，如数据采集算法、数据汇聚算法、数据治理算法、数据处理和分析算法等。

2. 商业应用（Business）：大数据应用一般体现在描述性分析、预测性分析或者决策性分析等，任何一种应用都是围绕某个具体场景展开的，因此大数据价值得以实现的一个重要前提，是找到一个合适的应用场景，该应用场景既直击需求痛点并有投资回报预期，又有数据积淀和 IT 建设基础。大数据在这个场景的应用，能够进一步内生和富集更多数据并因而形成数据闭环，就能进一步体现和实现大数据价值。

3. 算力（Computing Power）：所谓算力，指的是设备的计算能力，显然，对于大数据应用而言，更精准（复杂）的算法以及更高效的计算需求都需要强大的算力支撑，因此算力是大数据价值实现的基本保障。

4. 数据（Data）：数据是大数据价值实现的基础，因此必须首先解决诸如数据在哪以及如何从不同的数据源获取数据，并进行有效的富集、汇聚和深入加工等问题，从而为应用提供数据支撑或高级语义支撑。

2017 年，中国政府发布了"新一代人工智能规划"，明确提出了大数据智能这一概念，其内涵在于形成从数据到知识、从知识到智能的能力，打穿数据孤岛，形成链接多领域的知识中心，支撑新技术和新业态的跨界融合与创新服务。显然，大数据智能进一步明确了，在上述四个维度共同发力的基础上实现大数据价值的基本路径。

此外，笔者还想提请读者注意，数据的本质是行为主体（例如人）在进行各类活动（生活、工作、娱乐）时的行为、偏好等痕迹被业务系统（或互联网、物联网等）记录在各个服务器里。因此，数据本身暗含着行为主体的隐私，所以围绕大数据价值的实现，数据隐私与安全保护是一个不可回避的重要内容。这一点在《大数据》一书中也有专章介绍。2018年5月25日，欧盟《通用数据保护条例》（简称GDPR）正式实施，在全球范围内掀起了数据保护改革的浪潮。在数据全球化深入发展的当下，如何在后GDPR时代建立起一套数据保护法律体系，既符合国际最佳立法实践又符合本国国情，既能为个人权益、数字经济、国家安全等系列利益保驾护航，又能在国际数据规则制定中占有话语权，也成为各国政策制定者普遍关注的领域。

通读《大数据》一书，萦绕于笔者心中的问题是：大数据会让我们的生活变得更好吗？如何才能让我们的生活变得更好呢？在阅读本书的过程中，笔者能够感觉到作者在写作时力求做到"在不牺牲准确性的前提下，尽可能写得简单"，同时又能做到周全和清晰。因此，本书尤其适合对大数据有兴趣的初学者，其真正的意义在于：为你打开了一扇门，并言明了很多方向。

于南京九乡河

2020年7月12日

目 录

前言　**1**

致谢　**3**

第一章　数据爆炸　**1**

第二章　大数据为什么不一般？　**14**

第三章　大数据存储　**27**

第四章　大数据分析法　**45**

第五章　大数据与医学　**61**

第六章　大数据，大商务　**77**

第七章　大数据安全与斯诺登事件　**91**

第八章　大数据与社会　**106**

字节大小量表　**113**

小写英文字母 ASCII 码表　**114**

索引　**116**

英文原文　**123**

前　言

　　关于大数据的书籍通常会沦为以下两类：要么是对事物的实际运行方式不提供解释；要么是只适合研究生使用的高难度数学教科书。本书的写作目的与这两类都有所不同，可被视为第三类方案。它涉及的内容包括：大数据运作方式的基础知识；大数据如何改变我们周围的世界；它对我们的日常生活及商业世界的影响。

　　数据原本指文件和论文，也可能带有几张照片，但现代意义上数据的含义却远不止于此。社交网站每分钟都会以图像、视频和电影的形态生成大量数据。当我们输入地址和详细信用卡信息进行在线购物时也会创建数据。现在，我们正以几十年前无法想象的速度在收集和存储数据，但是，正如我们将在本书中看到的那样，新的数据分析技术正在将这些数据转换为有用的信息。在撰写本书时，我发现，如果不围绕大公司对大数据的收集、存储、分析和使用来进行写作，就无法在真正意义上讨论大数据。由于谷歌和亚马逊等公司在大数据领域的引领地位，因此我会不时提及它们。

本书的第一章首先向读者介绍了通常意义上的数据多样性，然后解释了数字时代如何改变我们定义数据的方式。通过讨论数据爆炸，顺便引入大数据的概念，其中涉及计算机科学、统计数据以及它们之间的接口。在第二章至第四章中，我使用了大量的图表来协助解释大数据所依赖的一些新方法。第二章探讨了大数据的特别所在，并借此提出了更具体的定义。在第三章中，我们讨论了与存储和管理大数据有关的问题。大多数人都熟谙在个人计算机上备份数据，然而，面对今时正在生成的海量数据，我们该如何应对？为了回答这个问题，我们将讨论数据库存储，以及在计算机集群之间分配任务的想法。第四章认为，大数据只有在我们可以从中提取有用信息时才有价值。通过简述几种公认的技术，我们可以了解如何将数据转换为有价值的信息。

然后，我们从第五章以大数据在医学中的作用开始，对大数据应用进行更详细的讨论。第六章通过亚马逊公司和奈飞公司（Netflix）的案例来分析商务实践，每个案例都彰显了使用大数据进行营销的不同特征。第七章探讨了围绕大数据的一些安全问题以及加密的重要性。数据盗窃已成为一个大问题，我们看到的是新闻中出现的一些案例，包括斯诺登和维基解密。本章最后说明了，网络犯罪是大数据不可回避的问题。在最后一章，即第八章中，我们会看到，通过先进的机器人及其在工作场所中的使用，大数据如何改变我们生活于其中的社会。本书结束于对未来智能家居和智能城市的思考。

本书只是简短的入门，不能论及大数据的所有内容，希望读 者能通过"进一步阅读"中的建议来继续自己感兴趣的探究。

致　谢

　　当我对彼得说要感谢他对本书的贡献时，他给了我如下建议："我要感谢彼得·哈珀，如果不是他执着于使用拼写检查器的话，这本书会完全不一样。"还有，我要感谢他在咖啡制作方面的专业知识和他的幽默感！这种支持本身就已无价，但彼得所做的远不止于此。可以说，没有他的持续鼓励和建设性贡献，这本书就不会写出来。

道恩·E.霍尔姆斯

2017年4月　xvii

第一章

数据爆炸

什么是数据？

公元前431年，斯巴达向雅典宣战。修昔底德在对战争的描述中，记载了被围困于城中的忠于雅典的普拉蒂亚部队，如何翻越由斯巴达领导的伯罗奔尼撒军队所建的围墙而最终得以逃脱的过程。要做到这一点，他们需要知道城墙的高度，以便制造高度合适的梯子。伯罗奔尼撒军队所建城墙的大部分都覆盖着粗糙的灰泥卵石，但他们最终还是找到了一处砖块清晰可见的区域。接下来，大量的士兵被赋予了一项任务，就是每个人分别去计数这些裸露砖块的层数。要在远离敌人攻击的距离之外完成判断，误差难以避免。但正如修昔底德所解释的那样，考虑到计数的是众多的个体，最常出现的那个数应该是可靠的。这个最常出现的数，我们今天称之为**众数**，普拉蒂亚人正是使用它来计量围墙的高度。由于使用的墙砖的大小是已知的，因此适合翻越城墙所需高度的梯子也顺理成章地打造了出来。随后，数百

1

的军人得以成功逃脱。此事可以被视为数据收集和分析最为生动的范例，也因而载入史册。但是，正如我们在本书的后续章节中将要看到的，数据的收集、存储和分析甚至比修昔底德的时代还要早几个世纪。

早在旧石器时代晚期的棍棒、石头和骨头上，人们就发现了凹口。这些凹口被认为是计数标记，尽管学术界对此仍然存有争议。也许最著名的例子是，1950年在刚果民主共和国发现的伊山戈骨，它距今大约有两万年之久。这个有着凹口的骨头被解读为具有特殊的功用，比如用作计算器或日历，当然也有人认为，骨头上的凹口只是为了方便手握。20世纪70年代在斯威士兰发现的列朋波骨甚至更为久远，时间大概可以追溯到公元前35000年左右。这块刻有29个线条的狒狒腓骨，与今天远在纳米比亚丛林中生活的土著仍然使用的日历棒，有着惊人的相似之处。这表明它确有可能是一种用来记录数据的方式，对于他们的文明来说，这些数据至关重要。

虽然对这些凹口骨骼的解释仍然没有定论，但我们清楚地知道，人类早期有充分记录的数据使用之一，是巴比伦人在公元前3800年进行的人口普查。该人口普查系统记录了人口数量和商品，比如牛奶和蜂蜜，以便提供计算税收所需的信息。早期的埃及人也擅长使用数据，他们用象形文字把数据写在木头或莎草纸上，用来记录货物的运送情况并追踪税收。但早期的数据使用示例，绝不仅限于欧洲和非洲。印加人和他们的南美洲前辈热衷于记录税收和商业用途的数据，他们使用一种被称为"奇普"的精巧而复杂的打彩色绳结的方法，作为十进制的记账系统。这些由染成明亮色彩的棉花或骆驼毛制成的打结绳，可以追溯到公元前3000年。虽然只有不到1 000个打结绳在西班牙人入侵和后续的各种毁灭性灾难中得以幸存，但它们是已知的第一批大规模数据存储系统的典范。现在有人正在开发计算机算法，试图解码"奇普"的全部含义，加深我们对其使用原理的理解。

虽然我们可以将这些早期的计数方法设想并描述为使用数据，但英文词data（数据）实际上是源于拉丁语的复数词，其单数形式为datum。今天，datum已经很少使用，"数据"的单数和复数都用data表示。《牛津英语词典》将该术语的第一个使用者，归于17世纪的英国神职人员亨利·哈蒙德。他在1648年出版的一本有争议的宗教小册子中使用了"数据"这个词。在此书中，哈蒙德在神学意义上使用了"数据堆"这一短语，来指称无可争辩的宗教真理。但是，尽管该出版物在英语中首次使用了"数据"这一术语，但它与现在表示"一个有意义的事实和数值总体"并不是同一个概念。我们现在所理解的"数据"，源于18世纪由普里斯特利、牛顿和拉瓦锡等知识巨人引领的科学革命。到1809年，在早期数学家的研究基础上，高斯和拉普拉斯为现代统计方法奠定了坚实的数学基础。

在更实际的层面上，当属1854年伦敦宽街暴发霍乱疫情时，针对该疫情收集的大量数据，它使得约翰·斯诺医生得以绘制了疫情图。数据和疫情图证明他先前的假设是正确的，即霍乱通过污染的水源传播，而不是一直以来被广为认同的空气传播。通过收集当地居民的数据，他确定患病的人都使用了相同的公共水泵。接下来，他说服地方当局关闭了该饮水源。关闭饮水源并不难，他们拆下了水泵的手柄，任务也就完成了。斯诺随后制作了一张疫情图,该图现在很出名,它清楚显示患病者以宽街的饮水泵为中心,成集群状态分布。斯诺继续在该领域潜心钻研,收集和分析数据,并成为著名的流行病学家。

约翰·斯诺之后,流行病学家和社会学家进一步发现,人口统计数据对于研究弥足珍贵。如今,在许多国家进行的人口普查,就是非常有价值的信息来源。例如,出生率和死亡率的数据,各种疾病的发生频率,以及收入和犯罪相关联的统计数据,现在都会有所收集,而在19世纪之前这些都是空白。人口普查在大多数国家每十年进行一次。由于收集到的数据越来越多,最终导致手工记录或以前使用的简单计数器,已经难以应对实际的海量数据登录。在为美国人口普查局工作期间,赫尔曼·何乐礼就遇到了如何应对这些不断增长的人口普查数据的挑战。

到1870年美国开展人口普查时,所依靠的是一种简单的计数器,但这种机器效率有限,已无法满足人口普查局的要求。1890年的人口普查有了突破,这完全得益于赫尔曼·何乐礼发明的用于存储和处理数据的打孔卡制表机。通常情况下,处理美国人口普查数据需要八年左右的时间,但使用这项新发明后,时间缩短到了一年。何乐礼的机器彻底改变了世界各国人口普查数据的分析处理,其中包括德国、俄罗斯、挪威和古巴。

何乐礼随后将他的机器卖给了一家后来称为国际商用机器(IBM)的公司,该公司开发并生产了一系列广泛使用的打孔卡机。1969年,美国国家标准协会制定了以何乐礼命名的打孔卡代码(或称何乐礼卡代码)标准,以对打孔卡机的先驱何乐礼表示敬意。

数字时代的数据

在计算机广泛使用之前，人口普查、科学实验或精心设计的抽样调查和调查问卷的数据都记录在纸上——这个过程费时且昂贵。数据收集只有在研究人员确定他们想要对实验或调查对象询问哪些问题后才能进行，收集到的这些高度结构化的数据按照有序的行和列转录到纸张上，然后通过传统的统计分析方法进行检验。到20世纪上半叶，有些数据开始被存储到计算机里，这有助于缓解部分劳动密集型工作的压力。但直到1989年万维网（或网络）的推出及其快速发展，以电子方式生成、收集、存储和分析数据才变得越来越可行。面对网络上可访问的海量数据，问题也接踵而来，它们需要及时得到处理。首先，让我们看看数据的不同类型。

我们从网络上获得的数据可以分为结构化数据、非结构化数据或半结构化数据。

手工编写并保存在笔记本或文件柜中的结构化数据，现在以电子的形式存储在电子表格或数据库中。电子表格样式的数据表由行和列组成，行记录的是数据，列对应的是字段（比如名称、地址和年龄）。当我们在线订购商品时，我们实际上也正在贡献结构化数据。精心构建和制表的数据相对容易管理，并且易于进行统计分析，实际上直到最近，统计分析方法也只能应用于结构化数据。

相比之下，像照片、视频、推文和文档这些非结构化数据就不太容易归类。一旦万维网的使用变得普遍，我们就会发现，很多这样的潜在信息仍然无法访问，因为它们缺乏现有分析技术所需的结构。但是，如果通过识别关键性特征，那么初看起来为非结构化的数据也可能不是完全没有结构。例如，电子邮件虽然正文的数据是非结构化的，但标题中包含了结构化**元数据**，因此它可以归类为半结构化数据。元数据标签本质上是描述性引用，可用于向非结构化数据添加可识别的结构化信息。给网站上的图像添加单词标签，它就可以被识别并且更易于搜索。在社交网站上也可以找到半结构化数据，这些网站使用主题标签，以便识别特定主题的消息（非结构化数据）。处理非结构化数据具有挑战性：由于无法将其存储在传统数据库或电子表格中，因此必须开发特殊工具来提取有用信息。在后面的章节中，我们会谈到非结构化数据的存储方式。

本章的题名"数据爆炸"一词，指的是逐渐产生的越来越多的结构化、非结构化和半结构化数据。接下来，我们将梳理产生这些数据的各种不同来源。

大数据简介

在本书的写作过程中，我在网上检索相关资料，体验了被网上可用的数据所淹没的感觉——来自网站、科学期刊和电子教科书的数据可谓海量。根据 IBM 公司最近进行的一项全球范围内的调查，每天产生的数据大约为 2.5 Eb。一个 Eb 是 10^{18}（1 后面跟 18 个 0）字节（或 100 万 Tb；请参阅本书结尾的"字节大小量表"）。在写作本书时，一台高配的笔记本电脑的硬盘通常会有 1 Tb 或 2 Tb 的存储容量。最初，"大数据"一词仅指数字时代产生的大量数据。这些海量数据（结构化和非结构化数据）包括电子邮件、普通网站和社交网站生成的所有网络数据。

世界上大约80%的数据是以文本、照片和图像等非结构化数据的形式存在，因此不适合传统的结构化数据分析方法。"大数据"现在不仅用于指代以电子方式生成和存储的数据总体，还用于指数据量大和复杂度高的特定数据集。为了从这些数据集中提取有用的信息，需要新的算法技术。这些大数据集来源差异很大，因此有必要让我们先详细了解一下主要的数据源以及它们生成的数据。

搜索引擎数据

到2015年，谷歌是全球最受欢迎的搜索引擎，微软的必应和雅虎搜索分居第二位和第三位。从谷歌可以查阅的最近一年数据来看，也就是2012年的公开数据，仅谷歌每天就有超过35亿次搜索。

在搜索引擎中输入关键词能生成与之最为相关的网站列表，同时也会收集到大量数据。网站跟踪继续生成大量数据。作为试验，我用"边境牧羊犬"为关键词进行了检索，并点击返回的最顶层网站。通过一些基本的追踪软件，我发现仅通过点击这一个网站就可以生成大约67个第三方站点的链接。商业企业之间通过此类方式共享信息，以达到收集网站访问者兴趣爱好的目的。

每次我们使用搜索引擎时，都会创建日志，它记录我们访问过的推荐网站。这些日志包含诸多有用信息，比如查询的术语、所用设备的IP地址、提交查询的时间、我们在各个网站停留的时长，以及我们访问它们的顺序——所有这些都以匿名的方式进行。此外，**点击流日志**记录了我们访问网站时所选择的路径，以及我们在网站内的具体导航。当我们在网上冲浪时，我们所做的每次点击都记录在某个地方以备将来使用。企业可以使用获取的软件来收集他们自家网站生成的点击流数据，这也是一种有价值的营销工具。通过提供有关系统使用情况的数据，日志有助于侦测身份盗用等恶意行为。日志还可用于评估在线广告的有效性，通过计算网站访问者点击广告的次数，广告的效用一目了然。

通过启用客户身份识别，"网络饼干"（Cookie）（一个小文本文件，通常由网站标识符和用户标识符组成）可用于个性化你的上网体验。当你首次访问所选网站时，"网络饼干"将被发送到你的计算机中，除非你已经禁用了它。以后每次你访问该网站时，"网络饼干"都会向网站发送一条消息，并借此跟踪你的访问。正如我们将在第六章中要看到的，"网络饼干"通常用于记录点击流数据，跟踪你的偏好，或将你的名字添加到定向广告中。

社交网站也会产生大量数据，脸书（Facebook）和推特（Twitter）位居榜首。到2016年年中，脸书平均每月有17.1亿个活跃用户。所有用户都在生成数据，仅日志数据每天就能达到大约1.5 Pb（或1 000 Tb）。视频共享网站优兔（YouTube）创建于2005年，目前广受欢迎，影响深远。在近期的新闻发布会上，优兔声称其全球用户数超过了10亿。搜索引擎和社交网站产生的有价值数据可用于其他许多领域，比如健康问题的处理。

医疗数据

如果我们看看医疗保健，就会发现一个涉及人口比例越来

越大的被电子化的领域。电子健康记录逐渐成为医院和手术的
标配，其主要目的是便于与其他医院和医生共享患者的数据，从
而提供更好的医疗保健服务。通过可穿戴或可植入传感器收集
的个人数据正日益增加。特别是为了健康监测，我们很多人都
在使用复杂程度各异的个人健身追踪器，它们输出前所未有的
新型数据。现在可以通过收集血压、脉搏和体温的实时数据，来
远程监控患者的健康状况，从而达到降低医疗成本并提高生活
质量的潜在目的。这些远程监控设备正变得越来越复杂，除了
测量基本生命体征参数之外，睡眠跟踪和动脉血氧饱和度也成
了测量的对象。

有一些公司通过激励措施来吸引员工使用可穿戴健身设
备，公司设定某些具体目标，比如减肥或每天走多少步路。作为
免费使用设备的条件，员工须同意与雇主共享数据。这似乎是
合理的，但不可避免地要涉及个人隐私。此外，选择加入此类计
划的员工很可能会承受额外的心理压力。

其他形式的员工监控也正变得越来越频繁，例如监控员工
在公司提供的计算机和智能手机上的所有活动。使用自定义软
件，此类监控可以包括从监视访问了哪些网站到记录键盘输入，
以及检查计算机是否用于私人目的（如访问社交网站）。在大
规模数据泄露的时代，安全性越来越受到关注，因此必须保护企
业数据。监控电子邮件和跟踪访问的网站，只是减少敏感资料
被盗的两种常用方法。

如前文所述，个人健康数据可以来自传感器，例如健身追踪
器或健康监测设备。然而，从传感器收集的大部分数据都以高
度专业化的医疗为目的。伴随着对各物种开展的基因研究和基

因组测序，产生了一批当今规模最为宏大的数据库。脱氧核糖核酸分子（DNA）以保存生物体遗传信息而闻名于世；1953年，詹姆斯·沃森和弗朗西斯·克里克首次将其描述为双螺旋结构。一个家喻户晓的基因研究项目是近年来的国际人类基因组计划，它的目标是确定人类DNA的30亿个碱基对的序列或确切顺序。这些数据最终会帮助研究团队进行基因疾病的探索。

实时数据

有些数据被实时收集、处理并使用。计算机处理能力的提高，惠及的不仅是数据处理，同时也大幅提升了数据生产能力。有时候，系统的响应时间至关重要，数据必须要得到及时处理。例如，全球定位系统（GPS）使用卫星系统扫描地球并发回大量实时数据。安装在你的汽车或内置在智能手机中的GPS接收设备，需要实时处理这些卫星信号才能计算你的位置、时间和速度。（"智能"表示某个物品，这里指的是手机，具有访问互联网的功能，并且能够提供可以链接在一起的多种服务或应用。）

该技术现在用于无人驾驶或自动驾驶车辆的开发。这样的车辆已经在工厂和农场等封闭的专门场所使用，一些大品牌汽车制造企业也在开发无人驾驶车辆，包括沃尔沃、特斯拉和日产等。相关的传感器和计算机程序必须实时处理数据，以便将车辆可靠地导航到目的地，并根据道路实况控制车辆的移动轨迹。这需要事先创建待行进路线的三维地图，因为传感器不能应对没有地图的路线。雷达传感器用于监控其他车流，并将数据发回控制汽车的外部中央执行计算机。传感器必须得到有效编程以探测不同的形状，并区分诸如跑进公路的孩子和风吹起的报纸这样的不同物体，或者甄别交通事故发生后的应急交通管制。然而，到目前为止，自动驾驶汽车还没有能力应对由瞬息万变的环境所带来的各种问题。

自动驾驶汽车首次致命碰撞事故发生在2016年。当时，驾驶员和自动驾驶仪都没有对切入汽车行进路线的车辆做出反应，也就是说没有任何制动的操作。自动驾驶汽车的制造商特斯拉在2016年6月的新闻稿中说，"引发事故的情况极为罕见"。自动驾驶系统会提醒驾驶员要始终将手放在方向盘上，并且还会检查他们是否在这样做。特斯拉表示，这是他们在1.3亿英里自动驾驶中发生的第一起死亡事故，而相比之下，美国每9 400万英里的常规驾驶（非自动驾驶）就会造成一人死亡。

据估计，每辆自动驾驶汽车每天平均生成30 Tb的数据，其中大部分数据必须立即处理。一个被称为**流计算**的新研究领域，绕过了传统的统计和数据处理方法，以期能提供处理这一特殊大数据的解决方案。

天文数据

2014年4月，国际数据公司（IDC）的一份报告估计，到2020年，数字世界将达到44万亿Gb（1 000 Mb等于1 Gb），数据总量 是2013年的十倍。天文望远镜所产生的数据与日俱增，例如位于智利的超大光学望远镜由四个望远镜组成，每晚都产生大量的数据，单个望远镜每晚所产生的数据就高达15 Tb。该望远镜在大型天气调查项目中起着引领的作用，它通过不停地扫描夜空制作和更新夜空图；该项目为期十年，产生的数据总量估计能达到60 Pb（2^{50}字节）。

在数据生成方面数量更大的是，建在澳大利亚和南非的平方公里阵列探路者（SKAP）射电望远镜。[①] 该望远镜预计于2018年开始运行。第一阶段它每秒将产生 160 Tb 的原始数据，随着建设进程的推进，产生的数据还会进一步的增加。当然，并非所有这些数据都会被存储，但即便如此，仍需要世界各地的超级计算机来分析剩余的数据。

数据到底有何用途？

如今我们的日常活动也会被收集并成为电子化的数据，想避免个人数据被收集几乎已经是不可能的事。超市收银机记录我们购买的商品的数据；购买机票时，航空公司收集我们旅行安排的信息；银行收集我们的财务数据。

大数据广泛应用于商业和医学，并在法律、社会学、市场营销、公共卫生和自然科学的所有领域得到运用。如果我们能够开发合适的数据挖掘方法，那么所有形式的数据都有可能提供大量有用的信息。融合传统统计学和计算机科学的新技术，使得分析大量数据变得越来越可行。统计学家和计算机科学家开发的这些技术和算法，可用以搜索数据模式。梳理出关键的模式，是大数据分析成功与否的关键。数字时代带来的变化大大改变了数据收集、存储和分析的方式。得益于大数据革命，我们才有了智能汽车和家庭监控。

以电子方式收集数据的能力，催生了令人兴奋的数据科学，也促成了统计学和计算机科学的融合。大量的数据得到有效分

---

① 原文中的ASKAP单指"澳大利亚平方公里阵列探路者"（Australian Square Kilometer Array Pathfinder），不包括南非的，可能是作者的笔误。——译注

析，从而在跨学科应用领域产生了新的见解，获得了新的知识。处理大数据的最终目的是提取有用的信息。例如，商业决策越来越依靠从大数据中分析所得的信息，并且期望值很高。但是，目前还有一些大难题亟待解决，尤其是缺乏训练有素的数据科学家，只有他们才能有效地开发和管理那些提取有用信息的系统。

通过使用源自统计学、计算机科学和人工智能的新方法，人们正在设计新的算法，有望推动科学的进步和产生新的科学见解。例如，尽管无法准确预测地震发生的时间和地点，但越来越多的机构正在使用卫星和地面传感器收集的数据来监测地震活动。其目的是想大致确定，从远期来看，可能会发生大地震的地方。美国地质调查局（USGS）是地震研究领域的主要参与者。该机构2016年预测："加利福尼亚州北部地区未来三十年发生里氏7级地震的概率为76%。"诸如此类的概率评估有助于将资源集中于重要事项，比如确保建筑物能够更好地抵御地震并实施灾害管理计划等。来自不同国家和地区的数家公司，正在使用大数据来改进地震的预测方法，这些方法在大数据出现之前是不可想象的。现在我们有必要来看一下大数据的非凡之处。 13

大数据为什么不一般？

　　大数据不是凭空而来的，它与计算机技术的发展密切相关。计算能力的快速提升和存储容量的迅猛增长，致使收集的数据越来越多。谁首创"大数据"这个术语现在已无从查考，但是它的本义一定与规模相关。然而，不可能仅根据生成和存储多少Pb甚或Eb来定义大数据。我们可以借由术语"小数据"来讨论由数据爆炸引发的"大数据"——尽管"小数据"并没有被统计学家广泛使用。大数据肯定是大而复杂的，但为了最终给出一个定义，我们首先需要了解"小数据"及其在统计分析中的作用。

大数据与小数据

　　1919年，罗纳德·费希尔来到位于英国的洛桑农业实验站分析农作物的数据。今天，费希尔被广泛认可为现代统计学这一强大学科的创始人。有关这些农作物的数据，来自19世纪40年代以来在洛桑进行的经典田间实验，包括针对冬小麦和春大麦所收集的数据，还有来自野外观测站的气象数据等。费希尔启动的项目被称为"实验田",目标是研究不同肥料对小麦的影响,目前该项目仍然在运行。

费希尔注意到,他所收集的数据颇为混乱,因此他将自己最初的工作称作"耙粪堆",这个说法后来变得很出名。然而,通过仔细分析研究那些记录在皮革装订的笔记本上的实验结果,费希尔终于理解了这些数据。没有充裕的时间,没有今天的计算技术,费希尔只有一个机械计算器。尽管如此,他还是成功地完成了对过去七十年累积的数据的计算。这个被称为"百万富翁"的计算器,依赖于单调乏味的手摇程序获取动力,但这在当时已经是创新的高科技了,因为它是第一个可以进行乘法运算的商用计算器。费希尔的工作是计算密集型的,没有"百万富翁"的帮助,他肯定无法完成计算工作。如果在今天,现代计算机在几秒钟内就能完成他所做的所有计算。

虽然费希尔整理并分析了很多数据,但今天来看,数据量并不算大,而且肯定不会被视为"大数据"。费希尔工作的关键是,使用精确定义和精心控制的实验,旨在生成高度结构化的、无偏的样本数据。鉴于当时可用的统计方法只能应用于结构化数据,这样做是必要的。实际上,这些宝贵的技术今天仍然是分析小型结构化数据集的基石。然而,我们今天可以使用的电子数据源是如此之多,以至这些技术已不再适用于我们现在可以访问的超大规模数据。

定义大数据

在数字时代,我们不再完全依赖于样本,因为我们经常可以收集到总体的所有数据。但是,这些越来越大的数据集的规

第二章　大数据为什么不一般?

15

模还不足以定义"大数据"——我们必须在定义中包含**复杂性**。我们现在处理的并非精心构建的"小数据"样本，而是不针对任何具体问题而收集的规模宏大的数据，它们通常都是非结构化的。为了描述大数据的关键特征，从而达到定义该术语的目的，道格·莱尼在2001年的文章中提出使用三个"v"来表征大数据：数量大（volume）、种类多（variety）和速度快（velocity）。通过依次审视这三个不同的"v"，我们就可以更好地了解"大数据"这个术语的含义。

数量大

"数量"指的是收集和存储的电子数据量，而且数据一直在持续的增加中。"大数据"一定很大，但到底有多大？以当前的眼光，给"大"设定一个数量标准是很容易的一件事，但我们应该明了，十年前被认为"大"的东西已经不再符合今天的标准。数据采集的增速是如此之快，任何设定的标准都将不可避免地很快过时。2012年，IBM公司和牛津大学报告了他们的大数据工作调查结果。在这项针对来自95个不同国家的1144名专业人士的调查中，超过一半的人认为1 Tb和1 Pb之间的数据集可视为"大"，然而有大约三分之一的受访者回答"不知道"。该调查要求受访者从八个选项中选择一个或两个表示大数据的特征，只有10%的人投票选择"数据量"，排名第一的选择是"范围广泛的数据"，该选项吸引了18%的人选。不能以"数量"门槛定义大数据还另有原因，比如存储和收集的数据类型这些因素，会随着时间的推移而发生变化并影响我们对数量的认知。诚然，一些数据集确实非常大，例如来自欧洲粒子物理研究所

（CERN）的大型强子对撞机的数据。它是世界上首屈一指的粒子加速器，自2008年以来一直在运行。即便只提取其总数据的1%，科学家每年需要分析处理的数据也会高达25 Pb。通常情况下，如果一个数据集大到不能使用传统的计算和统计方法进行 收集、存储和分析时，我们就可以说它满足了数量标准。像大型强子对撞机生成的这类传感器数据只是大数据的一种，所以让我们也看看其他类型的数据是何种情形。

种类多

虽然你可能经常看到"互联网"和"万维网"这两个术语被当作同义词而交替使用，但它们实际上是非常不同的概念。互联网是网络中的网络，由计算机终端、计算机网络、局域网（LAN）、卫星、手机和其他电子设备组成。它们都连在一起，通过IP协议从某个地址相互发送数据包。万维网（www或Web）的发明人伯纳斯—李将其描述为"全球信息系统"。在此系统中，互联网是一个平台，所有拥有联网计算机的个人都可以通过此平台与其他用户进行通信，比如通过电子邮件、即时消息、社交网络和短信进行交流。从互联网服务提供商（ISP）那里申请开通网络后，就可以获得"万维网"和许多其他服务。

一旦连接到万维网，我们就可以访问网络上那些无序而混杂的数据了。数据源既有可靠的，也有令人生疑的；重复和讹误的数据随处可见。这与传统统计所要求的干净和精确的数据相去甚远。尽管从万维网收集的数据有结构化的、非结构化的或半结构化的多种（例如社交网站上的文档或帖子等非结构化数据，电子表格等半结构化数据），但来自万维网的大数据主体上都是非结构化的。例如,全球的推特用户每天发布大约5亿条140个字符的消息或推文,这些数据都是非结构化的。[①] 推特上的这些短消息具有宝贵的商业价值,可以根据它们所表达的情绪划分为积极的、消极的和中立的三类。作为一个新领域,情感分析需要开发专门的技术。我们只有使用大数据分析法,才能有效地完成这项工作。虽然医院、军方和众多的商业企业出于各种目的,收集了大量差异化的数据,但从根本上说,它们都可以归类为结构化、非结构化或半结构化数据。

速度快

今天,万维网、智能手机和传感器等,正源源不断地生产着数据。速度自然与数量相关:生成数据的速度越快,数据量也就越大。例如,当今社交媒体上的消息常以滚雪球的方式传播,其传播方式与"病毒"无异。我在社交媒体上发布了某个内容,我的朋友们看到了,每个人都与朋友分享,朋友的朋友再发给朋友分享。很快,这些消息就会传遍世界各地。

速度也指数据被处理的速度。比如,传感器数据(像自动驾驶汽车生成的数据)必须实时生成。如果要确保汽车安全行驶,通过无线方式传送到数据中心的数据必须要得到及时分析,并将必要的指令实时发送回汽车。

可变性可以被认为是"速度"的附加维度,它指的是数据流量的变化率,例如高峰时段数据流量的显著增加。这一点也很重要,因为计算机系统在这个时段更容易出现故障。

---

① 北京时间2017年11月8日,推特将原先的发文长度上限从140个字符调整为280个字符。——译注

准确性

除了莱尼提议的三个基本的"v"（即数量大、种类多和速度快）之外，我们可以添加"准确性"为第四个维度。准确性是指所收集数据的质量。准确且可靠的数据，是20世纪统计分析的标志，为了设计出实现上述两个目标的方案，费希尔和其他一些学者可谓呕心沥血。但是，数字时代产生的数据通常是非结构化的，数据采集也常常在没有实验设计的前提下进行，甚至事先任何有价值问题的概念都没有。然而，我们就是寻求从这种大杂烩中获取信息。以社交网站生成的数据为例，这些数据本质上是不精确、不确定的，甚至通常被发布的信息就是彻头彻尾的谬误。那么，我们如何相信这些数据能产生有意义的结论呢？数量可以克服数据的如上缺陷。正如我们在第一章中所看到的那样，修昔底德所描述的普拉蒂亚部队让尽可能多的士兵计数砖块，就是想发挥数量的优势，以期获得他们意欲翻越的城墙的精确（或接近精确）高度。然而，我们需要多一个心眼，正如统计理论所告诉我们的，更大的数量会导致相反的结果。数据量越大，虚假的相关性就越多。

可视化和其他的"v"

在描绘大数据时，"v"不再固定，具有了可选择性，在莱尼最初的3v之外，竞争性的新词汇有"脆弱性"（vulnerability）和"可行性"（viability）等词，其中最重要的或许是"价值"（value）和"可视化"（visualization）。"价值"一般指的是大数据分析结果的质量。它也被用来描述商业数据企业对其他公司出售数据，而购买了数据的公司会利用自己的分析方法处理和使用数据，因此，"价值"是一个在数据商业领域中经常被提及的术语。

"可视化"虽然并不是大数据的特征，但其在展示数据分析结果和交流中意义重大。常见的静态饼图和柱状图已经得到进一步优化，它们之前用以帮助我们理解小数据集，现在也可以在大数据可视化方面发挥作用，但其适用性仍然有限。例如，信息图虽然能进行更复杂的数据呈现，但却是静态的。由于大数据是持续增加的，所以最佳的可视化手段应是用户交互式的，且创建者应能进行定期更新。比如，当我们使用GPS规划汽车旅行时，我们访问的是一个基于卫星数据的高度交互性图像，该图像能对我们进行定位。

综上所述，大数据的四个主要特征，即数量大、种类多、速度快和准确性，给数据管理带来了巨大挑战。我们在应对挑战的过程中期待获得的优势，以及我们期望用大数据来回答的问题，都可以在数据挖掘中找到答案。

大数据挖掘

在工商界和政界领袖中，"数据就是新石油"这句话广为流传。大家普遍认为，它是乐购（Tesco）客户忠诚卡的创始人克莱夫·胡姆比于2006年提出的。这句话不仅朗朗上口，也指出了数据的特征：它既像石油一样异常珍贵，但也必须先经过处理才能实现其价值。数据供应商们最初使用这句话作为营销口号，是为了销售自己的产品，他们试图让企业相信大数据就是未来。未来可期，但这个比喻于当下而言却并不完全准确。一旦你发现了油矿，你就拥有了适销对路的商品。但大数据不同，除非你有正确的数据，否则你无法创造任何价值。所有权是个问题；隐私也是个问题；而且与石油不同的是，数据似乎是一种无限资源。但是，也不必苛责这种以石油作比的说法，大数据挖掘的任务，确实是从大量的数据集中提取有用和有价值的信息。

利用数据挖掘、机器学习方法与算法，不仅可以侦测数据中的异常模式或异常现象，也可以进行预测。为了从大数据集中获得这类信息，我们可以使用有监督机器学习技术或无监督机器学习技术。有监督机器学习，有点类似于人类从例证中学习知识的过程。通过学习有正确范例标记的训练数据，计算机程序会生成规则或算法，然后据此对新数据进行分类。算法会通过测试数据进行验证。相较之下，无监督机器学习的算法，使用的是无标记的输入数据，且不给出数据处理目标，旨在探究数据并发现其中的隐藏模式。

我们可以将信用卡欺诈侦测作为一个例子，从而了解每种方法是如何工作的。

信用卡欺诈侦测

人们在侦测和防止信用卡欺诈方面做了很多努力。如果你不幸接到了信用卡欺诈侦测办公室的电话，你可能会好奇，他们是如何确定最近你卡上出现的消费很可能是欺诈消费的。由于信用卡交易的次数非常多，人们已无法再用传统数据分析技术通过人工来侦测交易活动，因此，大数据分析正变得越来越重要。金融机构不愿透露欺诈侦测的具体细节是可以理解的，因为这样做会让网络罪犯获知具体细节，从而找到规避欺诈侦测的方法。但是，粗线条的简单描述也能让我们感知其趣味所在。

　　信用卡欺诈存在几种可能的情形，我们可以先看看个人银行业务。假设是信用卡被盗，且诈骗者利用被盗信息，如PIN码（个人识别码）使用了信用卡。在这种情况下，信用卡支出可能骤增，发卡机构很容易就能侦测到这种欺诈行为。但更常见的情况是，诈骗者会先用盗取的信用卡进行"测试交易"，在"测试交易"中，他们会购买一些并不昂贵的商品。如果此次交易后平安无事，那么诈骗者接下来就会刷取更大的数额。这样的交易活动可能是欺诈，也可能不是，或许持卡人的购买模式发生了变化，也或许就是那个月花了很多钱而已。那么，我们如何侦测哪21　些交易是欺诈呢？首先来看一种被称为**聚类**的无监督技术，以及在上述情况下它的工作原理。

聚　类

　　基于人工智能算法，聚类算法可用于侦测客户购买行为中的异常。我们通过研究交易数据找出交易模式，并据此侦测任何异常或可疑情况，这些异常情况可能是欺诈，也可能不是。

　　信用卡公司采集了大量数据，并利用这些数据建立个人档案，显示客户的购买行为。然后，通过迭代（即重复进行同一运算以生成结果）程序以电子方式识别具有类似属性的个人档案，从而获得聚类。例如，可以根据有代表性意义的支出范围或位置信息、客户的最高支出限额或购买的商品种类来定义聚类，每种标准都会形成一个独立的聚类。

　　信用卡提供商收集的数据，并未标记交易是否为欺诈。我们的任务是将这些数据作为输入数据，并使用合适的算法对交易进行精确分类。为此,我们需要在输入的数据中找到相似性,从而进行分组或确定聚类。例如,我们可以根据消费金额、交易地点、物品种类或持卡人年龄对数据进行分组。当达成新交易时,系统将对该交易进行聚类识别,若新交易与该客户的现有聚类标识不同,则新交易会被视为可疑交易。即使新交易属于常见聚类,但若与聚类中心相距甚远,那么仍是可疑交易。

例如,一位住在帕萨迪纳八十三岁的老奶奶突然买了一辆豪华跑车,若这与她平时的购物习惯,如去杂货店和理发店不一致,则会被视为反常现象。所有像老奶奶购买豪华跑车这样的反常事件都值得进一步调查,而联系持卡人通常是调查的第一步。图1是说明上述情况的简单聚类图解。

聚类B是老奶奶平时的月支出,该聚类也包括月支出与她

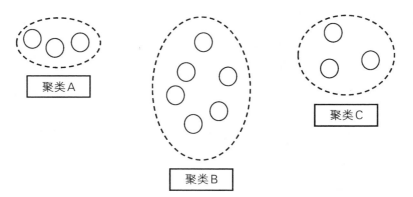

图1　聚类图解[①]

---

① 作者对表(table)和图(figure)没作区分,都使用了figure来表示,翻译时也统一使用"图"。——译注

类似的其他人。然而，在某些情况下，例如她的年度旅行期间，老奶奶的月支出就会有所增加，这样一来也许就会把她归到聚类C中。但是，聚类C离聚类B并不太远，所以并没有太大的不同。即便如此，由于这笔消费属于不同的聚类，也是可疑的账户活动，因而需要进行核实。像购买豪华跑车这样的行为则会被归于聚类A，这与她惯常所在的聚类B相去甚远，所以极有可能是非法交易。

与此相反的是，如果我们已经有了一组欺诈数据，我们就会使用分类法，而不是聚类算法，这是欺诈侦测中的另一种数据挖掘技术。

分　类

分类是一种有监督学习技术，前提是对相关群体事先就有所了解。我们以一个数据集为例，对该数据集的各种观察结果（先前掌握的知识）都已正确标记或分类。数据集被分为训练集和测试集两部分，训练集帮助我们构建数据的分类模型，测试集则用于检查分类模型是否良好。然后，我们就可以利用这个模型对新监测结果进行分类。

为了说明分类的具体情况，我们将构建一个用于侦测信用卡欺诈的小型决策树。

如图2所示，为了构建决策树，我们假设已经采集了信用卡交易数据，并且已根据所了解到的历史情况，将诸多交易划分为真实交易或欺诈交易。

通过这些数据，我们可以构建如图3所示的决策树，计算机可据此对输入系统的新交易进行分类。我们希望通过提出一系列问题,确定新交易的类别,即到底是真实交易还是欺诈交易。

信用卡是否被报遭窃或遗失？	商品购买是否异常？	被电告客户是否有购买行为？	分 类
否	否		真实交易
否	是	是	真实交易
否	是	否	欺诈交易
是			欺诈交易

图2　分类的欺诈数据集 24

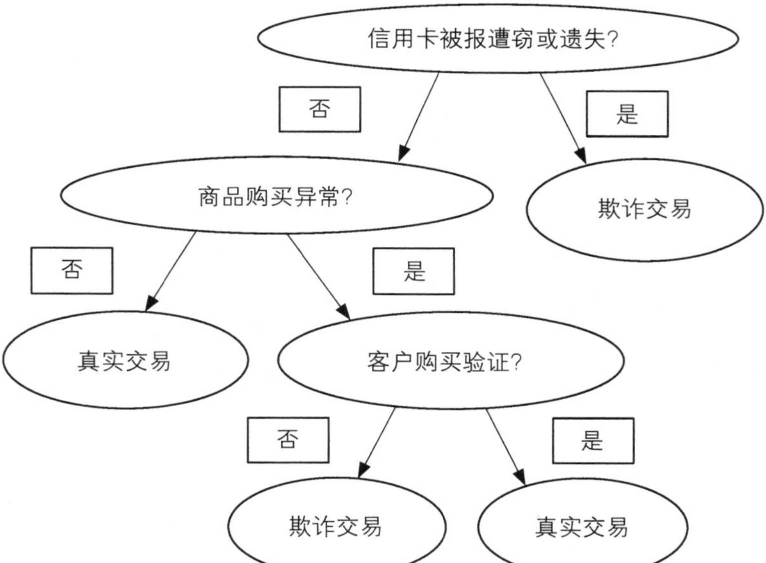

图3　交易决策树

如图3所示,从交易决策树的顶部开始,自上而下有一系列的测试问题,这些问题将帮助我们对新交易进行分类。

例如,如果史密斯先生的账户信息显示他已报告信用卡遗失或遭窃,那么任何使用此信用卡的交易都将被视为欺诈。若信用卡没有遗失或遭窃的报告,那么系统将会核查该客户是否购买了不寻常的商品,或是所购商品的金额是否符合客户的消费习惯。如果与往日没什么差别,那么这笔交易就会被视为正常交易,并被标记为"真实";反之,如果所购商品与往日相去甚远,银行则会致电史密斯先生。如果史密斯先生确认他购买了该商品,那么这笔交易则为真实交易,否则,此次交易就是欺诈。

　　在大致了解何为大数据,并讨论了大数据挖掘所能解决的问题之后,接下来让我们看看数据存储问题。

第三章

大数据存储

　　IBM 公司在加利福尼亚州圣何塞开发和销售的第一款硬盘驱动器的存储容量约为 5 Mb，存储在 50 个磁盘上，每个磁盘直径有 24 英寸。在 1956 年的时候，这绝对是尖端技术。作为大型计算机的组成部分，该硬盘驱动器体积庞大，重量超过一吨。到 1969 年"阿波罗 11"号登月时，美国宇航局在休斯敦的载人航天中心使用的大型计算机每台都有高达 8 Mb 的内存。令人惊讶的是，由尼尔·阿姆斯特朗驾驶的"阿波罗 11"号登月飞船的机载计算机只有 64 Kb 的内存。

　　计算机技术发展迅速。到 20 世纪 80 年代个人计算机热肇始之时，个人计算机硬盘驱动器的平均容量是 5 Mb。此时硬盘驱动器是选配的硬件，有些计算机就没有硬盘驱动器。今天，5 Mb 只够存储一两张图片或照片。计算机存储容量增长迅猛，虽然个人计算机存储量落后于大数据存储，但近年来也大幅增加。现在，你可以购买到配备 8 Tb 硬盘甚至更大的个人电脑。闪存现今的存储空间也达到了 1 Tb，足以存储约 500 小时的电影或超过30万张照片。这样的存储空间似乎已经很大，但当我们将其与每天产生的估计能达到2.5 Eb的新数据相比时，它们立刻就显得相形见绌。

一旦从真空管到晶体管的变化在20世纪60年代被触发后，放到芯片上的晶体管数量的增长就一发不可收。增长速度大致符合我们将在下一节讨论的摩尔定律。尽管有预测说小型化即将达到极限，但在到达极限前的进一步小型化仍然合理且价值连城。我们现在可以将数十亿个计算速度越来越快的晶体管植入同一块芯片，这样我们就可以存储更多的数据。与此同时，多核处理器和多线程软件也使得处理这些数据成为现实。

摩尔定律

1965年，英特尔的创始人之一戈登·摩尔曾预测道，在未来十年内，芯片中包含的晶体管数量将每二十四个月增加一倍。该预测在业内家喻户晓。1975年，他修改了自己的预测，认为芯片的复杂度将每十二个月翻一番并能持续五年，然后退回到每二十四个月增加一倍。摩尔的同事戴维·豪斯通过对晶体管增长速度的评估，认为微芯片的**性能**每十八个月就会增加一倍，这是当前摩尔定律最新的预测数值。事实证明，新的预测值非常准确。自1965年以来，计算机确实变得越来越快、越来越便宜、越来越强大，但摩尔本人认为这一"定律"很快就会失效。

根据米歇尔·沃尔德罗普2016年2月发表在科学期刊《自然》上的文章，摩尔定律的失效确实近在咫尺。微处理器是专门负责执行计算机程序指令的集成电路。它通常由数十亿个晶体管组成，晶体管嵌入硅微芯片的微小空间里。每个晶体管中的栅极能使它被接通或断开，用以存储0和1。极小的输入电流通过每个晶体管的栅极，并在栅极闭合时产生放大的输出电流。[27]米歇尔·沃尔德罗普对栅极间的距离尤感兴趣，目前顶级微处理器栅极的间隙是14纳米。他表示由于集成电路的进一步密集化，芯片过热以及如何有效散热等问题正制约着摩尔定律所预测的指数级增长。这也使我们注意到摩尔所言的迅速接近的基本限度。

纳米的尺寸是10^{-9}米或百万分之一毫米。拿其他物体来做个比较，人的头发直径约为7.5万纳米，原子的直径在0.1纳米到0.5纳米之间。供职于英特尔的保罗·加尔吉尼宣称，栅极间隙的极限为2纳米或3纳米。极限到来之日并不遥远——也许21世纪20年代就能见证。沃尔德罗普推断："在那样小的间隙之下，电子的行为将受控于量子不确定性，结果是晶体管变得无可救药般不可靠。"正如我们将在第七章中看到的那样，量子计算机（一种仍处于起步阶段的技术）有望最终能提供这一问题的解决方案。

摩尔定律现在也适用于数据的增长率，因为数据的增长似乎每两年大约翻一番。随着存储容量的扩大和数据处理能力的增强，数据也会增加。我们都是大数据的受益者。正是由于摩尔定律预测的成指数级增长的大数据，奈飞、智能手机、物联网（对连接到互联网的大量电子传感器所组成的万物互联的简便说法），以及云计算（将服务器连在一起的全球网络）等这一切才成为可能。所有这些新生事物生成的数据必须得到存储，下面我们将对此展开讨论。

结构化数据存储

每当我们使用个人计算机、笔记本电脑或智能手机的时候，
都是在访问存储在数据库中的数据。结构化数据（例如银行对
账单和电子地址簿）存储在关系数据库中。为了管理这些结构
化数据，需要使用关系数据库管理系统（RDBMS）来创建、维
护、访问和控制数据。第一步是设计数据库模式，即数据库的结
构。为了实现这一点，我们需要确定数据字段并将它们排列在
表中，然后我们再确定数据表之间的关系。一旦完成了上述工
作并构建数据库之后，我们就可以填充数据到数据库中，并使用
结构化查询语言（SQL）对其进行检索。

显然，数据表的设计必须要认真对待，否则后续的改动会大
费周章。但是，关系模型的价值不应被低估。对于很多结构化
数据应用程序来说，它快速且可靠。关系数据库设计的一个重
要考量被称为**规范化**，它包括最大化地降低数据的重复，从而降
低存储的压力。数据的访问会因此变得快捷，但即便如此，随着
数据量的增加，这种传统数据库性能的下降也不可改变。

问题主要关涉可扩展性。由于关系数据库基本上只能在一
台服务器上运行，所以随着数据库容量的逐渐增大，它就会变得
缓慢且不可靠。维持可扩展性的唯一方法是增加计算能力，但
计算力是有限度的。这被称为**垂直扩展性**。因此，虽然有关系
数据库管理系统存储和管理结构化数据，但是当数据过大时（如
Tb 或 Pb 及以上级），关系数据库管理系统就不再能有效工作，即
使对于结构化数据来说也是如此。

关系数据库的重要特性和持续使用它们的理由，在于它

们符合以下属性组，即原子性、一致性、独立性和持久性（简称ACID）。原子性确保不完整的处理（操作）无法更新数据库；一致性排除无效数据；独立性保证一个处理不会干扰另一个处理；持久性意味着数据库必须在执行下一个处理之前更新。所有这些都是理想的属性，但存储和访问大数据需要采用不同的方法，因为大数据基本都是非结构化的。

非结构化数据存储

对于非结构化数据来说，由于多种原因，关系数据库管理系统不再适用。关系数据库模型一旦构建完毕，就很难对其进行更改。这是甚为棘手的难题。此外，非结构化数据无法被方便地组织成行和列。正如我们所看到的，大数据通常是高速且实时生成的，需要得到实时处理。因此，尽管关系数据库管理系统使用范围广泛，而且也确实为我们提供了很好的服务，但鉴于目前数据的爆炸式增长，需要深入研究新的存储和管理技术。

为了存储海量数据集，数据被分配到不同的服务器上。随着所涉及的服务器数量的增加，出现故障的概率也随之增大。因此，将相同数据的多个副本存储在不同的服务器上就变得尤为重要。实际上，由于现在处理的数据量巨大，系统故障已经难以避免。数据存储已经内置了新的方法，以便应对这一难题。那么如何满足速度和可靠性（这两个相互矛盾）的双重需求呢？

海杜普分布式文件系统

分布式文件系统（DFS）为分布在多个节点的众多计算机上的大数据提供了高效且可靠的存储。谷歌公司于2003年10

月发表了一篇研究论文,该文是针对谷歌文件系统的推出而专门撰写的。在该论文的启发下,当时在雅虎工作的道格·卡廷和他的同事——华盛顿大学的研究生迈克·卡弗雷拉,开始了海杜普分布式文件系统的开发。海杜普是最受欢迎的分布式文件系统之一,它是一个名为海杜普生态系统的更大型开源软件项目的一部分。海杜普的命名取之于卡廷儿子的黄颜色大象软玩具,以流行的编程语言 Java 编写。在你使用脸书、推特或易贝(eBay)的时候,海杜普就会一直在后台运行。它不仅存储半结构化和非结构化数据,并且提供数据分析平台。

当我们使用海杜普分布式文件系统时,数据分布在许多节点上——通常是数万个节点,遍布于世界各地的数据中心。图4显示了单个海杜普分布式文件系统集群的基本结构,该集群由一个主管理节点和许多从属的数据节点组成。

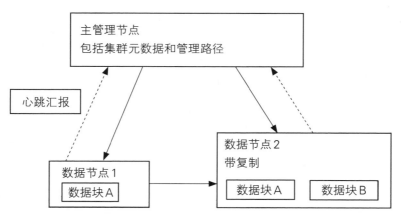

图4　海杜普分布式文件系统集群简略图

主管理节点处理来自客户端计算机的所有请求。它分配存储空间,并跟踪存储的有效性和数据的位置。它还管理基本文件操作（例如打开和关闭文件），并控制客户端计算机的数据访问。数据节点负责实际的数据存储，为此会根据需要创建、删除和复制数据块。

数据复制是海杜普分布式文件系统的基本特性。从图4中我们看到，数据块A同时存储在数据节点1和数据节点2中。重要的是，将数据块存储为多个副本，以保证在一个数据节点失效时，其他节点能够接管并继续处理任务而不至于丢失数据。为了跟踪哪些数据节点（如果有的话）已经失效，主管理节点每隔三秒会从各数据节点接收一条消息，也就是**网络心跳监测**。如果没有收到消息，则可推测所关涉的数据节点已经停止运行。因此，如果数据节点1无法发送心跳汇报，则数据节点2将成为操作数据块A的数据节点。如果主管理节点丢失，情况就不同了，在这种情况下，必须要使用内置备份系统。

数据只被写入数据节点一次，但应用程序将从中多次读取。每个数据块通常只有64 Mb，因此有很多很多的数据块。主管理节点的另一个功能是，确定在当前使用条件下哪一个为最佳数据节点，从而确保快速访问和处理数据。随后，客户端计算机就从所选出的最佳节点访问数据块。数据节点会随存储的需要适时添加，这一特征被称为**水平可扩展性**。

与关系数据库相比较，海杜普分布式文件系统的一个主要优点是：你可以收集大量数据，并不断添加数据，而且此时无须明了以后这些数据有何种用途。比如，脸书就使用海杜普来存储其不断增加的数据。任何数据都不会丢失，因为海杜普以原始格式存储所有的内容。根据需要添加数据节点很廉价，也不需要对现有节点进行更改。如果先前的节点变得冗余，也很容易让它们停止运行。正如我们所看到的,具有可识别的行和列的结构化数据存储在关系数据库管理系统中较为容易,而非结构化数据则可以使用分布式文件系统存储,这样做不仅廉价,而且方便。

32

用于大数据的非关系型数据库

NoSQL 是 Not Only SQL 的缩写,意为**"不仅是 SQL"**,它是非关系型数据库的统称。为什么需要一个不使用结构化查询语言的非关系型模型?我们可以简短地回答说:非关系型模型可以使我们不断添加新数据。非关系型模型具有管理大数据所必备的基本功能,即可扩展性、有效性和高性能。使用关系数据库,我们无法保证在不丧失功能的情况下进行数据的持续垂直扩展,而使用非关系型数据库,则可以通过水平扩展保持数据库的高性能。在描述非关系型分布式数据库基础架构,以及阐明其适用于大数据的原因之前,我们先要了解CAP定理。

CAP定理

2000年,美国加州大学伯克利分校计算机科学教授埃里克·布鲁尔提出了CAP定理,CAP分别指一致性(C)、可用性(A)和分区容错性(P)。对于分布式数据库系统来说,一致性要求同类节点中存储的数据相同。因此,在前文的图4中,数据节点1中的数据块A应该与数据节点2中的数据块A相同。可用性要求如果某个节点发生了故障,其他节点仍然能继续运行——如果数据节点1发生故障,数据节点2必须要继续运行。由于数据和数据节点分布在互为物理隔断的服务器上,这些机器之间的通信难免有时会失败。通信失败的情况被称为**网裂**。分区容错性要求，即便发生了这种情况，系统也要继续运行。

CAP定理的精髓在于，对于任何共享数据的分布式计算机系统来说，只能同时满足上述三个标准中的两个。因此，系统有如下三种可能性：具有一致性和可用性；具有一致性和分区容错性；或者具有分区容错性和可用性。请注意，由于在关系数据库管理系统中，网络未进行分区，因此只涉及一致性和可用性，并且关系数据库管理系统模型也同时满足这两个标准。在非关系型数据库中，由于分区是必然的存在，因此我们必须要在一致性和可用性之间做出选择。通过牺牲可用性，我们可以得到一致性。如果我们选择牺牲一致性，那么数据有时候会因服务器不同而有差异。

首字母缩略词BASE可以便捷地描述这种情况。这个缩略词有点像人为杜撰的，它指的是"基本可用（BA）、软状态（S）和最终一致（E）"。选择BASE的目的，是为了与关系数据库的ACID属性进行比照。在这里，"软状态"指的是对一致性要求的灵活性，可以允许有一段时间数据的不同步。最终目的是不放弃这三个标准中的任何一个，找到优化三者的方法，以达成妥协。

非关系型数据库的架构

由于结构化查询语言的无效，才有了非关系型数据库的诞生。例如，对于我们在图4中看到的连接查询，结构化查询语言是做不到的。非关系型数据库有四种主要类型：键值存储

数据库、列存储数据库、文档型数据库和图形数据库。它们对
于存储大型结构化和半结构化数据都非常有用。最简单的是
键值存储数据库，它由一个标识符（**键**）和与该键相关的数据
（**值**）组成，如图5所示。请注意，此处的"值"可以包含多个数
据项。

34

键	值
简·史密斯	地址：任何驱动器33号；任何城市
汤姆·布朗	性别：男；婚姻状况：已婚；子女数：2； 最喜欢的电影：《灰姑娘》《德古拉》《巴顿》

图5　键值存储数据库

　　当然，有很多这样的键值对，添加新键值或删除旧键值都非
常简单，这赋予了数据库高度的水平可扩展性。查询给定键的
值是数据库的主要功能。比如，使用键"简·史密斯"，我们可
以找到她的地址。对于大数据来说，键值存储数据库为存储提
供了快速、可靠且易于扩展的解决方案，但由于没有合适的查询
语言，它的优势受到了限制。列存储数据库和文档型数据库是
键值存储数据模型的扩展。

　　图形数据库遵循不同的模型，在社交网站中很受欢迎，并
且在商业应用程序中也能大显身手。这些图形通常非常大，特
别是用于社交网站的时候。在数据库中存储图形时，信息存储
在节点（即顶点）和边线中。例如，图6中的图形显示了五个节
点，它们之间的箭头表示关系。添加、更新或删除节点都会更改
图形。

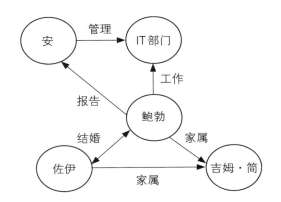

图6　图形数据库

在此示例中，节点是姓名或部门，边线反映的是它们之间的关系。通过查看边线中的关系实现从图形数据库中检索数据。例如，如果我想查询"IT部门中有子女的员工的姓名"，就会发现鲍勃同时满足这两个条件。值得注意的是，这不是有向图——我们不遵循箭头的指向，我们寻找的是关联。

目前，一种名为"新型数据库"（NewSQL）的检索方法正崭露头角。通过整合非关系型数据库的运作方式和关系型数据模型的ACID属性，这项潜在技术的目的是解决关系型数据模型存在的数据扩展问题，使其更适用于大数据。

云存储

像许多现代计算机术语一样，"云"（Cloud）听起来不仅友好、舒心、引人入胜，而且亲切熟悉。但如前文所述，"云"实际上只是相互连接的服务器所组成的网络的一种表达方式，它们分布在世界各地的数据中心。众多的数据中心连接在一起，构成了存储大数据的超级网络。

借助互联网，我们可以共享远程服务器上的应用。通过支付费用，就可以在由各公司提供的服务器上存储和管理我们的文件、运行应用程序等。只要你的计算机或其他设备具有访问云端的必备软件，你就能从世界上的任何地方查看文件，也可以授予其他人相同的权限。你还可以运行"驻留"在云端而不是本地计算机上的软件。这不仅是对互联网的简单访问，而且包括存储和处理信息，"云计算"因而得名。我们个人的云存储容量并不大，但积累起来的信息量则大得惊人。

亚马逊是最大的云服务提供商，但其管理的数据是商业秘
密。我们可以透过2017年2月亚马逊云存储系统S3**断网**（即服务丢失）这一事件来理解云计算的重要性。该断网事件持续了
将近五个小时，导致了众多网站和服务的连接中断，其中包括奈飞、亿客行（Expedia）和美国证券交易委员会。亚马逊后来在报告中说此事件是人为错误所致，他们的一名员工无意间切断了服务器的网络。重启这些大型系统的时间比预期的长了很多，但最终得以成功完成。不管怎么说，这一事件暴露了互联网的脆弱性，无论是真正的误操作，还是恶意的黑客入侵，都可以让其崩溃。

无损数据压缩

据广受尊重的国际数据公司2017年的估计，当时的数字世界总共有16 Zb的数据量，相当于不可思议的16×10^{21}字节。最终，随着数字世界的不断成长，诸如哪些数据应该保存，应该保留多少副本，以及需要保留多长时间等问题，我们都必须要面

对。定期清除数据或者对数据进行归档，这本身就是对大数据存在必要性的质疑。这样做不仅成本昂贵，并且可能会造成有价值数据的丢失，因为我们无法预知哪些数据在未来是重要的。随着存储的数据量的不断增加，为了最大限度地存储数据，数据压缩变得不可或缺。

由于收集到的电子数据质量良莠不齐，为要确保数据分析有效，在分析之前，必须要对其进行预处理，以便检查并解决一致性、重复性和可靠性方面的问题。如果我们要依赖从数据中提取的信息，那么一致性显然尤为重要。删除冗余的重复信息，对任何数据集来说都是良好数据管理的指标，而对于大数据，还有另外一层担忧，那就是，没有足够的存储空间来保存所有（重复的）数据。压缩数据可以减少视频和图像中的冗余，从而降低存储压力，而对视频来说，压缩数据可以提高流速率。

数字压缩分为无损和有损两种主要类型。在**无损压缩**中，所有数据都被保留，因此这对于文本特别有用。例如，扩展名为".ZIP"的文件已被压缩且没有丢失信息，因此对它们进行解压后，就会得到原始的文件。如果我们将字符串"aaaaabbbbbbbbbb"压缩为"5a10b"，就很容易明白压缩并还原成原始字符串的方法。压缩算法有很多，但先了解在不压缩的情况下数据是如何被存储的会大有裨益。

美国信息交换标准码（ASCII码）是对数据进行编码的标准方式，通过编码，数据才能够被存储在计算机中。每个字符都被指定为一个十进制数字，即ASCII码。正如我们已经知道的，数据存储为一系列的0和1。这些二进制数字被称为**比特**。标准ASCII码使用8比特（也定义为1个字节）来存储一个字符。例如，在ASCII

码中,字母"a"由十进制数97来表示,转换为二进制形式就是01100001。这些值能在标准ASCII表中查到,其中一小部分在本书末尾给出。大写字母具有与小写字母不同的ASCII码。

图7显示的是字符串"added"的编码情况。

字符串	a	d	d	e	d
ASCII	97	100	100	101	100
二进制	01100001	01100100	01100100	01100101	01100100

38 图7 编码的字符串

因此,"added"需要5个字节或40比特。基于图7的规则,使用ASCII码表进行解码。ASCII码显然不是编码和存储数据的经济方式,每个字符需要8比特空间似乎过多,并且也没有从字母使用频率有别这一角度来考虑问题,在文本文档中某些字母比其他字母的使用频率要高很多。有许多无损数据压缩模型,例如霍夫曼算法,它通过可变长度编码来节约存储空间。该技术就是基于字母出现的频率,用更短的代码表示那些出现频次最多的字母。

再以"added"为例,"a"出现一次,"e"也出现一次,"d"有三次。因为"d"出现的频次最高,它的编码应该最短。为了进行霍夫曼编码,我们首先计算"added"中各字母的次数:

$$1a \rightarrow 1e \rightarrow 3d$$

接下来,我们看到频次最小的两个字母分别为"a"和"e",用图8中的**二叉树**结构来表示。顶端的数字2是两个最小频次的字母出现次数相加的结果。

图8 二叉树   39

在图9中,新节点代表频次为3的字母d。

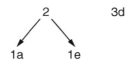

图9 有一个新节点的二叉树

图9是完整的树形图,各字母的频次位于顶端。在下面的图10中,树形图的边对应的编码是0或1,字母的编码就是沿边而上的数字。

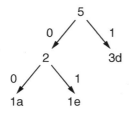

字母	码(位)
a	00
e	10
d	1

图10 完整的二叉树

因此,对于"added"来说,a=00,d=1,d=1,e=10,d=1,编码结果为0011101。使用该方法,存储字母"d"用了3比特,字母"a"和字母"e"都是2比特,总共是7比特。与原来的40比特相比,这是一个巨大的进步。

衡量压缩效率的方法之一,是使用数据压缩率,它指的是文件的实际大小与其压缩后大小的比率。在本例中,压缩率为

45/7，约等于6.43。高压缩率显示出良好的存储经济性。在实践中，这些数字之树会非常庞大，需要使用复杂的数学技术进行优化。此例展示了我们如何压缩数据而不会丢失原始文件所包含的任何信息，因此被称为无损压缩。

有损数据压缩

相比之下，声音和图像文件通常比文本文件大得多，因此使用了另一种被称为**有损压缩**的技术。之所以使用有损压缩，是因为当我们处理声音和图像时，无损压缩方法可能根本达不到足够高的压缩率，从而无法满足数据存储的需要。另一个同等重要的原因是，对于声音和图像来说，某些数据的丢失是可以容忍的。有损压缩通过永久删除原始文件中的某些无足轻重的数据，以达到减少存储空间的目的。此举的基本理念是，在不过度影响我们对图像或声音感知的前提下删除一些细节。

例如一张孩子在海边吃冰激凌的黑白照片，更准确地说是**灰度图像**，有损压缩的方法会从孩子的图像和大海的图像中删除同等量的数据。通过精确计算得出删除数据的百分比，确保观者对压缩后生成的图像在感觉上没有太大的差异——压缩率过大会导致照片模糊不清。在压缩率和图片质量之间需要权衡取舍。

如果要压缩灰度图像，我们首先将其分成8像素乘8像素的块。由于这是一个非常小的区域，因此所有像素的色调通常都相似。这一观察结果，以及有关我们如何感知图像的知识，对于有损压缩意义重大。每个像素都有一个介于0（纯黑色）和255（纯白色）之间的相应数值，数值表示灰阶度。在使用一种被称为离散余弦算法的方法进一步处理后，得到每个块的平均灰阶强度值，并将结果与给定块中的实际数值进行比较。由于平均值与实际值相差甚小，因此比较后的大多数结果为0或四舍五入后为0。有损算法将所有这些0收集起来，它们代表的是来自各像素的对图像不太重要的信息。在对这些在图像中高频出现的数值进行分组后，再使用一种被称为**量化**的技术对其中的冗余信息进行删除，从而实现数据压缩。例如，如果在64个值中（每 ₄₁个值需要1个字节的存储空间），我们有20个0，那么在压缩后，我们仅需要45个字节的存储空间。对构成图像的所有块都重复此过程，冗余信息会被悉数删除。

对于彩色图像，例如JPEG（联合图像专家小组格式），其算法识别红色、蓝色和绿色，并根据人类视觉感知的已知属性为每个色彩分配不同的权重。绿色的权重最大，因为人眼对绿色的感知强于对红色或蓝色的感知。彩色图像中的每个像素都分配有红色、蓝色和绿色权重，表现为〈R, G, B〉三元组。由于技术原因，〈R, G, B〉三元组通常转换为另一个三元组〈YCbCr〉，其中Y代表颜色的强度，Cb和Cr均为色度值，它们描述了实际的颜色。使用复杂的数学算法，可以减少每个像素的值，并通过降低保存的像素数量来最终实现有损压缩。

多媒体文件由于需要占用较大存储空间，通常都使用有损方法对其进行压缩。文件压缩得越多，再现质量就越差，但是由于牺牲了一些数据，因此可以实现很大的压缩率，从而使文件更小。

JPEG文件格式遵循的是联合图像专家小组于1992年首次制定的图像压缩的国际标准，它为压缩彩色和灰度照片提供了最流行的方法。目前该小组仍然非常活跃，每年都会开数次碰头会。

再以一张孩子在海边吃冰激凌的黑白照片为例。理想情况下，当我们压缩该图像时，我们希望照片中的孩子那部分保持清晰，为了实现这一点，我们可以牺牲背景细节的清晰度。美国加州大学洛杉矶分校亨利·萨穆利工程与应用科学学院的研究人员开发了一种被称为**数据扭曲压缩**的新方法，使上述想法成为现实。对此细节感兴趣的读者请参阅本书末尾的"进一步阅读"部分。

我们已经看到了分布式数据文件系统如何用于存储大数据。由于存储问题持续得到改善，现在大数据可用于回答以往我们无法回答的问题。正如下面我们将在第四章中所看到的那样，一种被称为**映射归约**的算法，可用于处理海杜普分布式文件系统中存储的数据。

第四章

大数据分析法

在讨论了如何收集和存储大数据之后，我们现在来看一看从数据中发现有用信息的技术，例如客户偏好或流行病传播速度等。随着数据集规模的增加，大数据分析法（相关技术的统称）的变化也日新月异，经典的统计方法为这种新范式提供了广阔的空间。

第三章介绍的海杜普提供了一种通过分布式文件系统存储大数据的方法。我们下面以映射归约（MapReduce）为例，来看一下大数据分析法。映射归约是一种分布式数据处理系统，它是海杜普生态系统核心功能的一部分。亚马逊、谷歌、脸书和许多其他组织，都在使用海杜普来存储和处理它们的数据。

映射归约

处理大数据的一种流行方法是将其分成小块，然后对它们分别进行处理。这基本上就是映射归约的工作方式。它将所需的计算或查询工作分配到众多计算机上来完成。通过去繁就简的例子来理解映射归约的工作方式非常值得尝试，实际情况也确实需要我们这样做，因为当我们手工操作时也需要依靠极为简化的示例，但它仍可演示用于大数据的分析过程。通常情况下，会有成千上万的处理器用来并行处理海量数据，可贵的是，该过程具有可伸缩性。这种想法不仅非常巧妙，且易于遵循。

映射归约分析模型由多个部分组成：**映射**（map）组件、**混洗**（shuffle）过程和**归约**（reduce）组件。映射组件由用户编写，并对我们感兴趣的数据进行排序。混洗过程是映射归约的核心，它根据键值对数据进行分组。最后是归约组件，它也由用户提供，通过对组值的合并得到最终的结果，并将其发送到海杜普分布式文件系统（HDFS）进行存储。

假设我们在海杜普分布式文件系统中存储了以下键值对文件，并具有下列各项的统计信息：麻疹、寨卡病毒、肺结核和埃博拉病毒。疾病是关键字（键名），值表示每种疾病的病例数。我们感兴趣的其实是各种疾病的总数量。

文件1：

麻疹, 3

寨卡, 2；肺结核, 1；麻疹, 1

寨卡, 3；埃博拉, 2

文件2：

麻疹, 4

寨卡, 2；肺结核, 1

文件3：

麻疹,3；寨卡,2

麻疹,4；寨卡,1；埃博拉,3

映射器可以让我们逐行分别读取各个输入文件,如图11所示。在读取数据之后,映射器将为每个不同的行返回键值对。

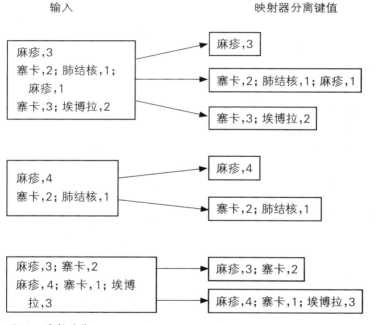

图11　映射功能

在对文件进行分片（split）并获取相应的键值对之后,算法的下一步是由主程序对键值进行排序和混洗。上述疾病将会按字母顺序排序,其结果被存入适当的文件以供归约程序调用,如图12所示。

第四章　大数据分析法

如图12所示，接下来，归约组件将映射和混洗阶段的结果组合起来，并将其（每种疾病）发送到各自独立的文件中。随后，算法中的归约步骤汇总各单位数据，并将汇总结果作为键值对发送至最终输出文件，该文件可以在分布式文件系统（DFS）中保存。

以上只是一个非常小的示例，实际上，映射归约方法可以帮助我们分析规模宏大的数据。例如，通过"爬网侠"数据平台，一家非营利机构提供的免费互联网副本，我们可以使用映射归约编写的计算机程序来计算每个单词在互联网上出现的频次。

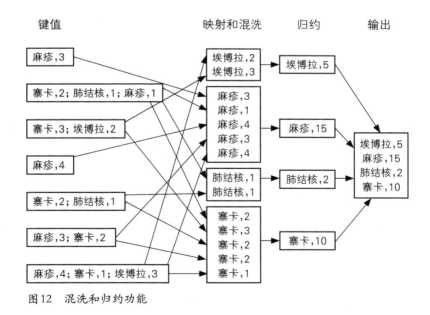

图12　混洗和归约功能

布隆过滤器

大数据挖掘的一种特别有用的方法是布隆过滤器，它是20

世纪70年代开发的一种基于概率论的技术。正如我们将要看到的那样，布隆过滤器特别适合于那些将存储作为主要考量、数据可存为列表的应用程序。

布隆过滤器背后的基本思想是，基于数据元素列表来构建一个系统，用以回答"列表中是否有 X ？"的问题。对于大数据集来说，搜索整个集合可能会费时太多而不具有实用性，因此布隆过滤器被当成了新的解决方案。该方法基于概率，并非100%准确，算法对某个元素是否属于列表进行判断，虽然会存在一定的误差，但它确实是一种从数据中提取有用知识的快速、可靠和利于有效存储的好方法。

布隆过滤器应用广泛。例如，它可用于检查特定的网络地址是否链接到恶意网站。在这种情况下，布隆过滤器充当的是已知恶意链接黑名单的角色，我们可以根据该黑名单快速准确地检查你刚刚单击的网络地址（URL）是否安全。新发现的恶意网址会被不断添加到黑名单中。由于现在有超过十亿个网站，而且每天都会有更多的网站诞生，因此跟踪恶意网站就成了一个应对大数据挑战的问题。

恶意电子邮件是类似的案例，它既可能是垃圾邮件，也可能包含恶意代码而具有钓鱼的企图。布隆过滤器为我们提供了一种甄别每个电子邮件地址的快速方法，如果发现地址存疑，就会及时发出警告。每个地址大约占用20个字节，对数目庞大的地址逐个存储和检查由于非常耗时而丧失实用价值，因为我们需要在极短的时间完成存储和检验的过程。通过使用布隆过滤器，可以显著减少存储的数据量。下面我们来构建一个小型的布隆过滤器，以了解它的工作原理。

假设我们有以下要标记为恶意电子邮件地址的列表：〈aaa@aaaa.com〉；〈bbb@nnnn.com〉；〈ccc@ff.com〉；〈dd@ggg.com〉。为了构建布隆过滤器，首先假定当前计算机上有10比特可用的内存。它就是所谓的**位数组**，起始值为空。对于一个位来说，只有两个状态，通常用0和1表示。因此，我们首先将位数组中的所有值都设置为0（表示空），而那些值为1的位，则意味着与其相关联的索引至少被分配过一次，这一点我们马上就能看到。

位数组的大小是固定的，无论我们添加多少内容，其大小都将保持不变。我们给位数组建立索引，如图13所示。

索引	0	1	2	3	4	5	6	7	8	9
位值	0	0	0	0	0	0	0	0	0	0

图13　10位数组

现在，我们需要引入**哈希函数**。哈希函数是一种算法，旨在将给定列表中的每个元素映射到数组中的某个位置。通过此举，我们现在仅需要将映射的位置存储在数组中，而不是电子邮件地址本身，这样存储空间就得以减少。

出于演示的目的，我们只展示了使用2个哈希函数的情形，但通常情况下我们会使用17个或18个函数，所涉及的数组也会大很多。由于这些函数被设计为均匀地映射，因此每次将哈希算法应用于不同的地址时，每个索引都有相等的机会被映射到。

那么，首先我们使用哈希算法将每个电子邮件地址分配给数组中的某个索引。

如果要将"aaa@aaaa.com"添加到数组中，首先要传递给哈希函数1，该函数将返回数组索引或位置值。例如，假设哈希函数1返回了索引3，被再次应用于"aaa@aaaa.com"的哈希函数2返回了索引4，那么这两个位置的存储位值都会变成1。如果某个设置的值已经为1，则保留原值，不作改动。同理，添加"bbb@nnnn.com"可能会导致位置2和7被占用或位值被设置为1；而"ccc@ff.com"可能会返回位置4和7。最后，假定哈希函数应用于"dd@ggg.com"并返回了位置2和6。图14是上述结果的汇总。

实际的布隆过滤器数组如图15所示，被占用位置的值设置 49 为1。

数　据	哈　希　1	哈　希　2
aaa@aaaa.com	3	4
bbb@nnnn.com	2	7
ccc@ff.com	4	7
dd@ggg.com	2	6

图14　哈希函数结果汇总

索引	0	1	2	3	4	5	6	7	8	9
位值	0	0	1	1	1	0	1	1	0	0

图15　侦测恶意电子邮件地址的布隆过滤器

那么，我们如何能将此数组作为布隆过滤器使用呢？假设现在我们收到了一封电子邮件，并且希望检查该电邮地址是否在恶意电子邮件地址列表中。假定该地址映射到位置2和7，那么它们的值都是1。由于返回的所有值都等于1，因此它很**可能**在列表中，因此也就很**可能**是恶意邮件。我们不能肯定地说它一定在该列表中，因为位置2和7也可能是其他地址的映射结果，并且索引也可能被多次使用。正因为如此，对元素所进行的列表登录与否的侦测还具有误报的概率。但是，如果任何一个哈希函数都返回了值为0的数组索引（别忘了，通常会有17个或18个函数），那么我们就能肯定该地址不在列表中。

虽然其中涉及的数学知识很复杂，但是我们可以看出，数组越大，未被占用的空间就越多，因而误报的次数或错误匹配的机会就越少。显然，数组的大小由所使用的键及哈希函数的数量来决定，但无论如何，它必须有足够大的数量以确保有充足的未被占用的空间，从而能使过滤器有效运行，并最大程度地减少误报的次数。

布隆过滤器反应速度很快，它可以非常有效地侦测欺诈性信用卡交易。过滤器会检查特定项目是否属于给定的列表或集合，异常交易会被标记为常规交易列表之外的行为。例如，如果你从未使用信用卡购买过登山装备，则布隆过滤器就会将购买攀岩绳的行为标记为"可疑"。另一方面，如果你确实购买过登山装备，则布隆过滤器就会将此次购买识别为"可接受"。当然事实也未必尽然，出错的概率依然存在。

布隆过滤器也可用于过滤垃圾邮件。垃圾邮件过滤器给我们提供了一个很好的范例，因为我们并不知道确切的所要查找的内容。通常情况下，我们寻找的只是模式，因此如果我们希望将包含"mouse"一词的电子邮件均视为垃圾邮件，那么我们同时就希望将包含诸如"m0use"和"mou\$e"之类变体的邮件也当作垃圾邮件。事实上,我们希望将包含该词所有可能的可辨别变体的邮件都识别为垃圾邮件。过滤与给定单词不匹配的所有内容是非常容易的,因此我们将只允许"mouse"通过过滤器。

布隆过滤器还可用于提升网络查询结果排名算法的速度,这是致力于网站推广的人士颇为感兴趣的话题。

佩奇排名

当我们使用谷歌搜索时,返回的网站会依据其与搜索词的相关性进行排名。谷歌主要通过一种被称为"佩奇排名"(PageRank)的算法来实现排序。人们普遍认为,这个算法是以谷歌的创始人之一拉里·佩奇的姓氏来命名的。他与公司的联合创始人谢尔盖·布林合作,发表了有关该新算法的论文。在 2016年夏季之前,佩奇排名的结果可以公开获取,只要下载使用佩奇排名工具条就可以得到结果。公开的佩奇排名工具条指标范围从1到10。在该工具被下架之前,我保存了一些结果。如果我使用笔记本电脑在谷歌搜索栏中输入"大数据",则会得到这样一条消息:"大约3.7亿个结果(0.44秒)",佩奇排名指标为9。页面网页列表的顶端是一些付费广告,随后是维基百科。使用"数据"为关键词进行检索,返回的结果是:约55.3亿个结果(0.43秒),佩奇排名指标为9。其他的都是佩奇排名指标为10的例子,其中包括美国政府网站、脸书、推特和欧洲大学协会。

佩奇排名的算法基于指向网页的链接数——链接越多,得分越高,页面作为搜索结果的显示就越靠前。佩奇排名与访问页面的次数多少无关。如果你是网站设计师,你一定想优化你的网站，以使它能在给定的某些关键词搜索时靠近列表的顶部，因为大多数人只会看前三个或四个搜索结果。这需要大量的链接，因此链接交易在业内就成了公开的秘密。为了打击"人工"排名，谷歌会分配一个新的0排名给有牵连的公司，甚至将它们从谷歌完全删除，但这并不能解决问题；交易只是被迫潜入地下，链接继续被出售。

佩奇排名本身并没有被废弃，它现在是一个大型排名程序的一部分，只不过它不再提供给公众查看。谷歌会定期重新计算排名，以便及时反映新链接和新网站的情况。由于佩奇排名具有商业敏感性，因此无法获取其详细而完整的资料，但是我们可以通过一个示例来了解它的总体思路。佩奇排名算法是基于概率论所提出的一种分析网页之间链接的复杂方法，其中概率1表示"确定性"，概率0表示"不可能"，其他所有结果的概率值都位于二者之间。

要搞清楚排名的工作原理，我们首先需要知道什么是概率分布。如果我们投一个六面匀称的骰子，那么从1到6这六个数字的概率是相等的，也就是说每个数字的概率都为1/6。所有可能结果的汇总以及与之相关的概率就是概率分布。

回到我们按照重要性对网页进行排名的问题，我们不能说每个网页都同等重要，但是如果我们有一种能为每个网页分配概率的方法，则可以合理地表示网页的重要性。因此，诸如佩奇排名之类的算法所要做的就是为整个网络构建概率分布。为了解释这一点，让我们设想有一个随机的网络浏览者，他实际上可能从任何网页开始，然后通过有效的链接进入另一个页面。

假设有一个简单的网络，它只有三个网页，分别为"大数据1"、"大数据2"和"大数据3"。页面间只有从"大数据2"到"大数据3"，从"大数据2"到"大数据1"，以及从"大数据1"到"大数据3"的链接。该网络的结构如图16所示，其中节点代表网页，箭头（边缘）表示链接。

每个页面都有一个佩奇排名，代表其重要性或受欢迎程度。"大数据3"的排名最高，因为指向它的链接最多，因此点击率也最高。假如现在那个随机浏览者访问了一个网页，如果我们把他或她对下一个网页的浏览视为投票，那么对所有备选的下一个网页来说得票的概率是均等的。例如，如果我们的随机浏览者当前正在访问"大数据1"，则接下来的唯一选择是访问"大数据3"。因此，可以说"大数据1"对"大数据3"投了1票。

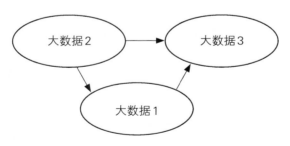

图16　小型网络的有向图

在真实的网络中，新链接会不断涌现。因此，假设我们现在发现"大数据3"链接到"大数据2"，如图17所示，则"大数据2"的佩奇排名将发生变化，因为随机浏览者在浏览"大数据3"之后有了一个可供选择的网页继续浏览。

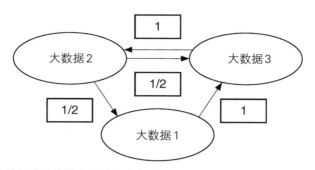

图17 增加链接后的小型网络有向图

　　如果我们的随机浏览者从"大数据1"开始，那么接下来的唯一选择就是访问"大数据3"，因此"大数据3"得到了1张票，也是全部的票。相比之下，"大数据2"的得票数为0。如果他或她从"大数据2"开始，则投票数被平均分配至"大数据3"和"大数据1"。最后，如果随机浏览者从"大数据3"开始，则他或她的全部投票只能投给"大数据2"。图18是对上述投票方式的汇总。

　　从图18我们可以看到每个网页的总得票数如下：

　　"大数据1"的总票数是1/2（来自"大数据2"）

　　"大数据2"的总票数是1（来自"大数据3"）

　　"大数据3"的总票数是1½（来自"大数据1"和"大数据2"）

	大数据1的投票比率	大数据2的投票比率	大数据3的投票比率
支持大数据1	0	1/2	0
支持大数据2	0	0	1
支持大数据3	1	1/2	0

图18 各网页的得票数

由于冲浪者起始页的选择是随机的,因此起始页的机会均等,它们初始的佩奇排名分配值都为1/3。为了给以上示例最终赋值,我们需要根据每个页面的得票数更新初始的佩奇排名。

例如,"大数据1"仅从"大数据2"那里得到了1/2票,因此"大数据1"的佩奇排名为$1/3 \times 1/2 = 1/6$。与之类似,"大数据2"的佩奇排名为$1/3 \times 1 = 2/6$,"大数据3"的佩奇排名为$1/3 \times 3/2 = 3/6$。由于所有网页佩奇排名的总数值为1,因此我们就得到了一个概率分布,它可以显示各网页的重要性或排名情况。

但是实际情况要复杂一些。我们说过,随机浏览者选择任意网页的概率为1/3。第一步之后,我们计算出随机浏览者浏览"大数据1"的概率为1/6。那么,第二步之后呢?好了,我们再次使用当前的佩奇排名作为得票数来计算新的佩奇排名。此轮的计算略有不同,因为当前的佩奇排名不相等,但是方法大同小异。计算所得的新的佩奇排名如下:"大数据1"为2/12,"大数据2"为6/12,"大数据3"为4/12。重复这些步骤或迭代,直到算法收敛为止,也就是说,对佩奇排名的计算过程一直持续,直到无法通过进一步的乘法运算修改数值为止。得出最终排名后,佩奇排名就可以为给定搜索选择排名最高的页面。

佩奇和布林在其原始研究论文中提出了一种计算佩奇排名的方程式,其中包括一个阻尼因子d,它表示随机浏览者单击当前页面上任一链接的概率。因此,随机浏览者不单击当前页面上任一链接的概率为(1-d),也意味着该随机浏览者已经完成了浏览。正是阻尼因子确保了经过足够数量的迭代计算后,整个网络上的平均佩奇排名值稳定为1。佩奇和布林在论文中说,经过52次迭代后,包含3.22亿个链接的网络佩奇排名会趋于稳定。

公共数据集

有许多免费的大数据集,感兴趣的团体或个人可以将其用于自己的项目。本章前面提到的"爬网侠"就是其中一例。作为亚马逊公共数据集的一部分,到2016年10月的时候,"爬网侠"存档的网页数超过了32.5亿个。公共数据集包含广泛的专业领域数据,包括基因组数据、卫星图像和全球新闻数据。对于不太可能自己编写代码的人来说,谷歌的"N元浏览器"(Ngram Viewer)提供了一种有趣的方式来交互式浏览一些大型数据集(有关详细信息,请参阅"进一步阅读")。

大数据范式

我们已经知道了大数据的一些有用之处,在前面的第二章我们还讨论了小数据。对于小数据分析来说,科学方法是行之有效的,并且必然涉及人机互动:某人有了个想法,提出了假设或模型,并设计了测试真伪的方法。著名的统计学家乔治·博克斯在1978年写道:"所有模型都是错误的,但有些却是有用的。"他的意思是,一般而言,统计和科学模型不能准确描述我们所处的世界,但是好的模型也是有用的,我们可以据此进行预测并自信地得出结论。但是,正如我们已经看到的,在处理大数据时,我们并不遵循这种方法。相反,我们发现处于主导地位的是机器,而不是科学家。

托马斯·库恩在1962年的著作中描述了"科学革命"的概念,它发生在规范科学的现有范式在相当长的时间内得到充分

56

发展和研究之后。当难以解决的异常现象不断出现,现有的理论受到挑战,研究人员会对理论失去信心,此时可以说"危机"来了。"危机"最终将由新的理论或范式来化解。如果新的范式要想被人们接受,那么它必须能够解答旧范式不能应对的一些问题。但是,总的来说,新范式不会完全压倒以前的范式。例如,从牛顿力学到爱因斯坦相对论的转变,改变了科学界看待世界的方式,但并没有使牛顿定律过时:牛顿力学变成了范围更广的相对论的一个特例。从经典统计学到大数据分析的转变,也代表了巨大的变化,并且具有范式转变的诸多特征。因此,不可避免地需要开发新的技术来应对这种新情况。

下面讨论一下在数据中寻找相关性的技术,该技术通过变量之间的关系强度进行预测。经典统计学已经确认,相关并不意味着因果关系。例如,老师有可能记录了学生的缺勤数和成绩,然后,老师发现两者之间存在明显的相关性,他或她可能会使用缺勤数来预测成绩。然而,缺勤会导致成绩差这个结论是错误的。仅通过盲目计算的结果,无法知道为什么两个变量之间具有相关关系:也许学习能力较弱的学生具有逃课的倾向;也许由于疾病而缺勤的学生以后无法追赶。只有通过对数据的分析和揣摩,才能确定哪些相关性是真实有用的。

至于大数据,使用相关性会产生更多的问题。如果我们使用一个庞大的数据集,编写的算法会返回大量的虚假相关,它们与任何人的见解、观点或假设都大相径庭。错误的相关会产生问题,例如离婚率和人造黄油消费之间的关系,这只是媒体报道的许多虚假相关中的一例。通过应用科学方法,我们可以看到这种所谓的相关性原来如此荒谬。实际上,随着变量的增加,虚假相关的数量也会上升。这是试图从大数据中提取有用知识面临的主要难题之一，因为在用大数据挖掘这样做的时候，我们通常寻求的就是模式和相关性。正如我们将在第五章中看到的那样，谷歌流感趋势预测失败的原因之一，就是这些问题。

58

第五章

大数据与医学

大数据正在显著地改变医疗保健行业，但其潜能尚未被充分认识。它的价值在包括医疗诊断、流行病预测、衡量公众对政府健康警示的反应，乃至减少与医疗保健体系有关的支出等领域，都有待重新评估。我们从**保健信息学**开始讨论。

保健信息学

运用前几章中描述的通用技术，可以对医疗大数据进行收集、存储和分析。广义上说，保健信息学及其多个分支学科，如临床信息学和生物信息学，利用大数据改进病人护理和降低成本。我们来看看大数据的定义标准（第二章中已讨论）——数量大、种类多、速度快和准确性，以及这些标准如何适用于医疗数据。例如，通过从社交网站收集与公共卫生相关的数据来跟踪流行病，数量大和速度快这两项都能得到满足；种类多也能达到要求，由于病历是以文本格式保存的，结构化和非结构化的都有，此外，收集的数据还包括如核磁共振提供的传感数据；准确性是医学应用的基础,需特别小心以便消除不准确数据。

　　通过从脸书、推特等网站,各种博客、留言板以及网络搜索等收集到的数据,社交媒体成为医疗相关信息的潜在宝贵源泉。关注具体医疗保健问题的留言板随处可见,提供了大量非结构化数据。使用类似于第四章中描述的分类方法,采集脸书和推特上的帖子以监测药物不良反应经历,可以为医疗保健专业人士提供有关药物相互作用和药物滥用的重要信息。采集社交媒体数据用于公共卫生相关研究,现在也是学术界公认的做法。

　　塞尔莫情报(Sermo Intelligence)是一家全球医疗网站,它自称是"全球最大的保健数据收集者"。像这种面向医学专业人士的社交网站,为保健人士提供了与同事互动的机会,从而产生即时的众包福利。在线医疗咨询网站越来越受欢迎,并将生成更多信息。尽管收集到的庞大的电子病历不能被公开访问,但或许是最重要的信息源。这些病历,通常根据其首字母被称为EHR,它是病人完整病史的电子版,包括诊断、处方用药、医学影像(如X光),以及全程收集的所有其他相关信息。据此可以构建起"虚拟病人"——我们将在本章下文中讨论这一概念。在使用大数据改进病人护理和降低成本的同时,通过汇集各种在线渠道生成的信息,也使得对新暴发的流行病进程的预测成为可能。

谷歌流感趋势

　　与许多国家一样,美国每年都会遭受流感袭击,导致医疗资源紧张和许多人失去生命。公共健康监测机构,即美国疾病控制中心提供的过往流行病数据,加上大数据分析,为研究人员预测疾病的传播，进而实施精准服务和减少疾病的影响提供了强大支撑。

谷歌流感趋势团队是利用搜索引擎数据进行流感预测的先行者。他们感兴趣的是：如何在每年的流感预测时能走在美国疾病控制中心的前面。2009年2月出版的权威科学杂志《科学》上发表了一封信，由六位谷歌软件工程师组成的团队解释了他们正在从事的工作。如果数据能够用来预测每年流感的进程，那么疾病就能得到及时遏制，从而拯救生命并节省医疗资源。为了实现这一目标，谷歌团队的做法是收集和分析与流感相关的搜索引擎查询数据。以往利用在线数据预测流感暴发的尝试要么失败，要么收效甚微。不过，通过学习这些早期的开拓性研究所犯的错误，谷歌和美国疾病控制中心有望能成功利用搜索引擎查询生成的大数据来实现对流行病的跟踪。

美国疾病控制中心与欧洲的类似机构——欧洲流感监督计划（EISS），从各种渠道收集数据，其中包括内科医生报告的类似流感症状的病人数量。核实数据通常需要两周左右的时间，而在此期间，流感又会进一步扩散。运用从互联网上实时收集的数据，由谷歌和美国疾病控制中心组成的团队旨在提高流行病预测的准确性，并且每二十四小时发布一次预测结果。为了做到这一点，从流感相关搜索查询中收集的数据，覆盖从关于流感药方和症状的单个互联网搜索，到诸如打给医疗咨询中心的电话这样的大量数据。谷歌能够利用其从2003年到2008年间的大量搜索查询数据，并通过IP地址能够确定搜索查询的发生地，从而按照州别对数据进行分组。美国疾病控制中心的数据收集自十个地区，每一地区包含出自数个州的累积数据（例如第

61

九区包括亚利桑那州、加利福尼亚州、夏威夷州以及内华达州），数据最终被集成到模型中。

谷歌流感趋势计划基于如下假设：有关流感的在线搜索数量与去医生诊所的数量高度相关。假如某个特定地区有很多人在线搜索有关流感的信息，那么就可以预测流感病大概率会扩散至邻近区域。由于我们的兴趣是寻找趋势，数据可以是匿名的，因而无须征得个人同意。运用五年来与美国疾病控制中心在同一时段，即流感季节所收集的数据，谷歌对覆盖所有主题的共5000万条最常见的搜索查询逐条统计，并计算出各搜索词每周出现的次数。然后，将这些搜索查询统计结果与美国疾病控制中心的流感数据进行比较，使用相关性最高的数据建构流感趋势模型。谷歌选用了前45个与流感相关的搜索关键词，并进而在人们的搜索查询中进行跟踪。完整的搜索关键词表属于秘密，包括诸如"流感并发症""流感治疗""流感一般症状"等。以历史数据为基准，评估搜索关键词与流感活动的相关性，再将新的实时数据与这一数据做比对，这样就建立起了从一级到五级的分类，五级为最严重级别。

在2011—2012年及2012—2013年美国流感季节，谷歌的大数据算法未能成功预测。这一事件引起广泛关注。流感季节结束后，他们将自己的预测与美国疾病控制中心的实际数据做了对照。基于现有数据所建立的模型本该能表征流感趋势，但实际情况是，谷歌流感趋势算法在其被运用的年份里，对流感病例数量的预测高出了至少50%。模型失灵有多种原因。有些搜索项被刻意排除，只是因为它们与研究团队的预期不符。广为人知的例子是高中篮球，它与流感看似毫不相干，却与美国疾病控制中心的数据高度拟合,但被排除在模型之外。变量选取,即选取最恰当预测项的过程,始终是一个具有挑战性的工作,只有通过优化算法才能避免偏差。谷歌将其算法细节定为秘密,仅透露出高中篮球在搜索关键词表的前100位中,并辩解说将其排除是有理由的,因为流感和篮球在一年中同一时间达到峰值只是虚假相关。

我们已经注意到,谷歌在模型构建中运用了45个搜索项作为流感搜索因子。假如他们仅仅使用一个搜索项,如"流行性感冒"或"流感",那么诸如"感冒药"之类的所有搜索将被忽略。数量充足的搜索项会提高预测的精准度,但搜索项过多也会降低预测的精度。运用当前数据作为训练数据来构建模型,可以预测未来数据的走向。但是如果预测因子太多,训练数据中那些微不足道的随机病例就会被模型化。如此一来,模型尽管与训练数据非常拟合,却不能很好地进行预测。这种看似自相矛盾的现象,叫作"过拟合",谷歌团队对此没有充分关注。因其与流感季节巧合而忽略高中篮球还情有可原,但是对于5 000万个不同的搜索项来说,数量太过庞大,不可避免地会出现与美国疾病控制中心数据高度相关,而与流感趋势并不相关的搜索项。

类似流感症状的病人去看医生,诊断结论常常不是流感(例如只是普通感冒)。谷歌运用的数据从搜索引擎查询中选取收集而来,由于数据收集出现了偏差,比如将不使用计算机的和使用其他搜索引擎的人排除在外,因而得到的结果并不很科学。 63 预测结果不佳可能还有另一个原因,在谷歌上搜索"流感症状"的用户很可能浏览过多个与流感相关的网站,导致同一搜索被多次计算,从而使数字被夸大。另外,搜索行为因时而异,特别是在疾病流行期间，这一点也需要予以考虑并定期更新模型。预测中一旦出现错误，就会产生连锁反应。谷歌流感预测中发生的正是这种情况：某一周出现的错误传递至下一周。搜索查询按实际是否发生而被收集和分析，并没有按拼写和措辞进行分组。谷歌自己就有将"流感的症候""流感症候""流感之症候"分别计入的先例。

谷歌的这项始于2007—2008年间的工程受到大量批评。这些批评有时并不公正。诟病的主要对象是缺乏透明度，比如拒绝透露全部搜索选项，不愿回应学术界获取信息的请求。搜索引擎查询数据并非精心设计的统计实验产品，如何找到一种方法对这些数据进行有意义的分析以获取有用知识，这是一个全新的和富有挑战性的领域。然而，这也需要合作。2012—2013年的流感季节，谷歌算法发生了重大变化，开始运用一种被称为弹性网络的相对较新的数学技术，此举为选取必要的和减少不必要的预测因子提供了严谨的方法。2011年，谷歌启动了一个跟踪登革热的类似项目，但他们不再发布预测。2015年，谷歌流感趋势被撤销。不过，现在他们开始与学术科研人员分享他们的数据了。

谷歌流感趋势是运用大数据进行流行病预测的早期尝试之一，它为后来的研究人员提供了灵感。尽管其结果未达预期，但将来有可能开发出新技术，将大数据用于跟踪流行病的潜能充分释放。美国洛斯—阿拉莫斯国家实验室的一群科学家，就运用维基百科的数据做过此类尝试。卡内基·梅隆大学德尔菲研究小组，在美国疾病控制中心"流感预测"的竞赛中拔得头筹——在2014—2015年和2015—2016年两个年度均为最精确

64

的预测者。研究小组运用来自谷歌、推特以及维基百科的数据成功监测了流感的暴发。

西部非洲埃博拉暴发

我们的世界,过去经历过很多流行病。1918—1919年的西班牙流感,死亡人数在2 000万到5 000万之间,共有大约5亿人感染。由于对病毒知之甚少,没有有效治疗方法,公共卫生响应非常有限——因对疾病不明就里,此是必然。1948年,负责通过全球合作协同监督和改进全球卫生状况的世界卫生组织(WHO)成立,此种窘境才得以改变。2014年8月8日,在国际卫生条例紧急委员会的电视电话会议上,世界卫生组织宣布:西部非洲埃博拉病毒的暴发正式构成了"国际关注的突发公共卫生事件"(PHEIC)。世界卫生组织明确指出,埃博拉的暴发是一个"非常事件",需要国际社会付出史无前例的努力加以遏制,从而避免疾病的大流行。

2014年西部非洲埃博拉的暴发,主要限于几内亚、塞拉利昂和利比里亚。这与美国每年一度的流感暴发有所不同,因此也提出了一系列不同的问题。埃博拉的历史数据要么无法获得,要么毫无用处,因为此等规模的暴发从未有过记录,制定新的应对策略迫在眉睫。考虑到人口流动数据对公共卫生专业人士监督流行病的扩散会有帮助,人们认为移动电话公司掌握的信息可以用来跟踪感染区的人员行踪,如果再加上其他措施,例如旅游限制,就可以遏制病毒传播,并最终拯救生命。据此构建的疾病暴发实时模型,会预测出下一个病例很可能会出现的区域,然后再针对性地聚集资源。

从移动手机上可以获取基本的数字信息，如呼叫者和被呼叫者的电话号码、呼叫者的大致方位等。移动手机呼叫都会留下痕迹，根据呼叫使用的发射塔，可以大致判断呼叫者的位置。接触这些数据也产生一些问题：隐私问题首当其冲，电话被追踪的人身份完全暴露，而他们自己对此却一无所知。

在受埃博拉影响的西部非洲国家，移动手机分布密度并不均衡，在贫穷的乡村地区比例最低。例如2013年的时候，在利比里亚和塞拉利昂，拥有手机的家庭刚刚过半，它们是2014年疾病暴发时受到直接影响的两个国家。但即便如此，它们仍然能够提供足够多的数据用于跟踪人口流动情况。

一些重要的移动电话数据交给了弗洛明德基金会。这是一家总部设在瑞典的非营利机构，致力于使用大数据从事面向世界最贫穷国家的公共卫生服务工作。2008年，弗洛明德基金会率先运用移动运营商的数据跟踪医学难以应对的人口流动情况。这是世卫组织消灭疟疾倡议的一部分，于是该基金会就成了应对埃博拉危机的首选机构。另一个著名的国际团队运用重要的匿名数据，构建了埃博拉感染地区的人口流动地图。由于流行病蔓延期间人们的行为与往日不同，这些数据用途有限，但对于人们遇到紧急情况会倾向去往何地，也给出了强烈的暗示。移动电话信号塔的记录提供了实时人口活动的详情。

不过，世卫组织发布的埃博拉预测数字比实际记录的病例要高出50%以上。

谷歌流感趋势与埃博拉预测具有类似的不足，即运用的预测算法仅仅基于初始数据，未将动态变化考虑在其中。实际上，这些模型均假定病例数量未来会继续上升，上升速度与医学干

66

预开始之前并无二致。显然，医疗和公共卫生措施会产生预期的正面效果，而这些并没有被融入模型之中。

由伊蚊传播的寨卡病毒于1947年在乌干达被首次记录，此后病毒传播远至亚洲和美洲。最近的寨卡病毒暴发于2015年发生在巴西，它导致了又一起国际关注的突发公共卫生事件。对于大数据统计建模，谷歌流感趋势以及埃博拉暴发期间的工作，已经给了我们很多教训。现在公认的是，数据应该从多种渠道收集。回想一下谷歌流感趋势计划仅从其自身的搜索引擎收集数据吧。

尼泊尔地震

那么，运用大数据进行流行病跟踪的未来前景如何？移动电话通话详细记录具有实时性特征，已经被用于协助监控灾难发生期间的人口流动情况，比如运用于范围广泛的尼泊尔地震区和墨西哥猪流感暴发区域。2015年4月25日，尼泊尔地震发生之后，由弗洛明德基金会牵头，汇集了来自英国的南安普敦大学和牛津大学，以及美国和中国的多个机构的科学家组成了一个国际团队，他们使用移动电话通话详细记录对人口流动情况做了评估。尼泊尔人持有手机的比例很高，利用1 200万用户的匿名数据，弗洛明德团队能够跟踪地震发生九天内人口的流动情况。这种快速反应部分是由于跟尼泊尔主要服务提供商有约在先，碰巧的是，合作细节在灾难发生前一周刚刚完成。提供商的数据中心拥有硬驱达20 Tb的专用服务器，这使得团队能够立即开展工作，从而能在短时间内让灾难救援机构获取地震发生九天内的信息。

大数据与智能医学

只要病人去医生办公室或是医院,相关电子数据按照惯例都要被收集。电子健康病历成为病人保健联络的法定文件:病人的病史、处方用药、检验结果均记录在案。电子病历还可能包括诸如核磁共振扫描等传感数据。数据匿名入库,供研究之用。据估计,在美国,普通医院平均存储超过 600 Tb 的数据,其中大部分为非结构化数据。如何挖掘这一数据以提供改进病人护理和减少成本呢?简单地说,我们将数据拿来,包括结构化和非结构化的,确定与病人相关的特征,运用诸如分类和回归等统计技术将结果建模。病人数据主要是非结构化文本格式,要有效进行分析需要使用如 IBM 公司的"沃森"人工智能所使用的那种自然语言加工技术。这将在下一节讨论。

据 IBM 公司预计,到 2020 年,医学数据每七十三天会翻一番。可穿戴设备被越来越多地用以监测健康个体,广泛使用于计算我们每天行走的步数,测量和平衡我们需要多少卡路里,跟踪我们的睡眠模式,以及给出我们心率和血压的即时信息等。获得的这些信息,可被上传到我们的个人电脑。记录由我们私下保存,或者——有时会有这种情况——自愿与雇主共享。这种关于个体的真实的数据级联,为医疗保健专业人士提供了有价值的公共卫生数据,并提供了一种识别个体变化的方法,这些变化可能对避免诸如心脏疾病等的突发有所裨益。与人群相关的数据,有助于医生根据患者的特征来跟踪监测特定治疗方案的副作用。

2003 年,人类基因组计划完成之后,基因数据会逐渐成为我们个人医疗记录的重要组成部分，它本身也为研究提供了海量数据。人类基因组计划的目的是绘制人类基因图谱。总的来说，有机体的基因信息被称为基因组。典型的人类基因组包含大约两万组基因，绘制这样一个基因组需要大约100 Gb的数据。绘制基因图谱是一个高度复杂、高度专业和多元化的领域，但运用大数据分析法的意义令人神往。收集到的基因信息存储在庞大的数据库中，近来一直有人担心这些数据会遭到黑客攻击，并将捐献DNA的病人查找出来。有人建议，为安全起见，应该往数据库中添加虚假信息，当然虚假数据的量应控制在一定范围之内，以防数据库变得对医学研究毫无价值。由于需要管理和分析基因组学产生的大数据，生物信息学这一跨学科领域应运而生。近年来，基因测序速度越来越快，成本越来越低，绘制个人基因组图谱现在已切实可行。算上十五年的研究成本，第一个人类基因组测序花费将近300万美元。而现在，很多公司能以合理的价格给个人提供基因组测序服务。

　　诞生于人类基因组计划的虚拟生理人类（VPH）计划，使用实际病人的庞大数据库来建立一套计算机仿真模型，让临床医生模拟看病治疗，找出特定病人的最佳治疗方案。将这些方案与类似症状及其他相关医疗细节进行比较，计算机模型可以预测出一个病人治疗的可能结果。进一步运用数据挖掘技术并结合计算机仿真，可找到个性化治疗方案。因此，诸如核磁共振这样的检查结果，可能会集成到仿真系统当中。未来的数字病人会包含真实病人的所有信息，并根据智能设备数据予以更新。不过，数据安全越来越成为该计划不得不面对的重要挑战。

医学中的沃森

2007年,IBM公司决定建造一台计算机,用来挑战美国电视游戏节目《危险边缘》中的顶级竞赛选手。以IBM公司的创始人托马斯·J.沃森名字命名的大数据分析系统应运而生。与之对垒的是两位《危险边缘》节目昔日的冠军:一位是布拉德·鲁特,74次参赛连胜;另一位是肯·詹宁斯,总共赢取了让人震惊的325万美元奖金。《危险边缘》是一档智力竞赛节目,节目主持人给出"答案",而参赛选手要猜出"问题"。参赛选手有三位,答案或线索出自多种类别,如科学、体育和世界历史等,也包括不太规范、有些奇怪的类别,诸如"之前和之后"。举个例子,对于线索"他葬在汉普郡教堂的墓地,墓碑上写着'骑士、爱国者、医生和文人;1859年5月22日—1930年7月7日'",正确答案是:"阿瑟·柯南·道尔爵士是谁?"有个比较边缘的类别叫"抓住这些人",对于"通缉犯,波士顿人,犯有19次谋杀,1995年潜逃,最后于2011年在圣莫妮卡海滩被抓获"这条线索,正确答案是:"白佬·巴尔杰是谁?"线索以文本格式发送给沃森,竞赛中略去音视频线索。

在人工智能领域,自然语言处理(NLP)被公认为是对计算机科学的巨大挑战,它对于沃森的发展最为重要。信息必须可访问、可检索,这在机器学习中是个难题。研究团队按照词汇应答类型(LAT)从分析《危险边缘》线索入手,将线索中确定的答案分类。在上面的第二个例子中,词汇应答类型是"这位波士顿人"。第一个例子中没有词汇应答类型,无法对代词"it"进行归类。IBM团队分析了两万条线索,找出了2 500个特有的词汇应答类型，但它们也只覆盖了大约一半的线索。接下来，解析线索以确定关键词以及它们之间的关系。再对电脑里相关的结构化和非结构化数据进行检索，然后基于初步分析提出假设，最后通过寻找更深层的证据，提出可能的答案。

要赢得《危险边缘》游戏，快速而先进的自然语言处理技术、机器学习、统计分析至关重要。其他要考虑的因素包括准确性和类别的选择。运用以往获胜者的数据，计算出合格表现的基准。几次尝试过后，整合了很多人工智能技术的深度问答分析给出答案。这一系统使用了多台计算机进行平行运算，但不连接互联网；它的计算基于概率和专家提供的知识。除了生成答案，沃森运用置信度评分算法凸显最佳结果。只有达到置信度阈值时，沃森才准备显示它已得出的答案，相当于参赛选手按响抢答铃。沃森战胜了两位《危险边缘》的冠军。詹宁斯坦然接受失败，用他本人的话说："我本人欢迎我们的计算机新霸主。"

以《危险边缘》的人工智能为基础，沃森医疗系统得以成功开发，它可以检索和分析结构化和非结构化数据。由于建立了自己的知识库，该系统本质上如同对特定领域人类思想过程的模拟。医学诊断是基于现有医学知识的，它依靠证据，要求输入的信息准确无误并包含全部相关信息，且具有一致性。人类医生拥有经验但会犯错误，诊断水平良莠不齐。沃森医疗系统的诊断过程与《危险边缘》中的人工智能相似，在通盘考虑全部相关信息后给出判断，每个判断都附有置信度等级。沃森内置的人工智能技术使它能够加工大数据，包括医学影像生成的海量数据。

现在，沃森超级计算机已经发展为多应用系统，亦是巨大的商业成功。另外，沃森已应用在人道主义工作中，例如通过特别研发的开源分析系统，帮助跟踪埃博拉在塞拉利昂的扩散情况。

医疗大数据的隐私问题

大数据显然具有预测疾病流行和定制个性化医疗方案的作用，但是，硬币的另一面——个人医疗数据的隐私又当如何应对呢？尤其是在当下，随着越来越多地使用可穿戴设备和智能手机应用，问题就产生了：数据归谁所有？该存储在哪里？谁可以接触和使用数据？面对网络攻击，安全如何保障？还有大量道德和法律问题都悬而未决。

健康跟踪器数据或许会让雇主得到，并用来做以下事情：好的方面，例如给达到某些指标的员工发奖金；不好的方面，确定哪些员工未能达标，兴许还会招致辞退。2016年9月，德国达姆施塔特工业大学和意大利帕多瓦大学的科学家合作研究团队，发布了他们对健康跟踪器数据安全的研究报告。让人震惊的是，在测试的17种不同厂家生产的健康跟踪器中，在防止数据被更改方面没有一家是安全可靠的。在保护数据准确性方面，仅有四家采取了措施，但这些保护措施都被该团队成功瓦解。

2016年9月里约奥运会之后，国家兴奋剂计划被曝光，多数俄罗斯运动员被禁赛。顶尖运动员，包括威廉姆斯姐妹、西蒙·拜尔斯和克里斯·弗鲁姆等人的医疗记录遭到黑客攻击，并被一群俄罗斯网络黑客在奇幻熊网站公开曝光。这些由世界反兴奋剂机构掌管在其数据管理系统ADAMS上的相关医疗记录，仅仅显示医疗用途豁免，并没有关于这些遭到网络中伤运动员的违禁行为。非法侵入系统可能源自鱼叉式网络钓鱼邮件，利用该技术，邮件被伪装成来自机构内部的高级可信源，如健康保健提供者，发给下级的信件。通过下载的恶意软件，该技术被用来非法获取诸如密码和账号等敏感信息。

防止大数据医疗数据库遭受网络攻击，进而确保病人的隐私越来越受到重视。匿名个人的医疗数据买卖是合法的，但即便如此，有时单个病人的真实身份还是有可能被发现。哈佛大学数据隐私实验室的科学家拉坦娅·斯威妮和刘吉素做了一个有价值的实验，显示了安全数据的脆弱性。他们运用合法获取的来自韩国的**加密**（即打乱排列顺序而使文件难以识读，见第七章）医疗数据，不仅能够解密医疗记录中的独特标识符，而且通过与公开的医疗记录进行比对就可以确定单个病人的身份。

医疗记录对网络罪犯极具价值。2015年，健康保险公司安森保险声称其数据库被非法侵入，超过7 000万人受到影响。分众识别的关键数据，如姓名、地址以及社会保障号码等被黑客组织攻破。他们运用被盗密码进入系统并安装恶意木马软件。尤为严重的是，在美国作为唯一身份证明的社会保障号码是不加密的，这给盗取身份留有极大便利。很多安全漏洞都是从人为错误开始的：人们太忙，注意不到网址中的细微差别。闪存盘等设备丢失、被盗，有时甚至一旦某个毫无戒心的员工将设备插入USB端口，设备瞬间就被蓄意植入恶意软件。心怀不满的员工和犯错的员工，都会导致不计其数的数据泄露。

世界知名机构，诸如美国的梅奥医学中心和约翰·霍普金斯医学院，英国的国民医疗服务系统（NHS），以及法国的克莱蒙费朗大学医院，都在加速将大数据这种全新的利好运用到医疗保健管理中。云系统让受权用户得以使用世界任何地方的数据。仅举一个例子,英国的国民医疗服务系统计划到2018年的时候,让医疗记录可以通过智能手机悉数获取。这些发展不可避免地会令他们使用的数据招来更多的攻击,因此需要加倍努力,开发出有效的安全方法以确保数据安全。

74

大数据

大数据，大商务

20世纪20年代，以"街角房子"咖啡馆闻名的英国餐饮提供商里昂公司，雇用了年轻的剑桥大学数学家约翰·西蒙斯做统计工作。1947年，雷蒙德·汤普森和奥利弗·斯坦汀福德双双被西蒙斯招募，派往美国做实情调查。正是这次美国之行，他们了解到电子计算机及其执行常规运算的能力。西蒙斯对他们了解到的情况非常重视，他设法说服里昂公司购买了一台计算机。

莫里斯·威尔克斯当时正在剑桥大学致力于建造电子延迟存储自动计算器（EDSAC）。在他的协助下，LEO计算机成功建成。该计算机依赖穿孔卡片运行，1951年首次用于基本会计事务，诸如将一列数字相加。到1954年，里昂公司已经形成了自己的计算机业务，并且正在建造"LEO II"系列，接着又建造了"LEO III"系列。尽管第一批办公计算机早在20世纪50年代就在安装使用，但由于它们使用电子管（"LEO I"系列是6 000个）和磁带，加上内存太小的缺陷，这些早期的机器并不可靠，应用也非常有限。最初的 LEO 计算机被普遍看作第一台商业化计算机，它为现代电子商务铺平了道路。经过数次合并，里昂公司于 1968 年成为新组建的国际计算机有限公司（ICL）的一部分。

电子商务

　　LEO 计算机和稍后出现的大型计算机，仅适合于诸如会计和审计之类的数字运算任务。传统上花费大量时间来统计数字列的工人，现在却要将时间花在制作打孔卡上，这不仅是一项烦琐的工作，同时还需要同样高的准确性。

　　由于在商业企业使用计算机已经可行，如何使用计算机提高效率，降低成本并收获利润，成了人们感兴趣的话题。晶体管的发展及其在市售计算机中的使用促成了机器的小型化，以至于在 20 世纪 70 年代初就有了建造个人计算机的想法。但是，直到 1981 年，IBM 公司在市场上推出 IBM-PC 并使用软盘进行数据存储时，建造个人计算机的想法才真正开始为商家所认可。个人计算机的文字处理和电子表格功能，很大程度上减轻了办公室烦琐的日常工作。

　　用软盘存储电子数据的技术很快让人们想到，将来无须使用纸张即可有效地开展业务。1975 年，美国《商务周刊》杂志发表的一篇文章推测，到 1990 年差不多会实现无纸化办公。理由是停止或显著减少纸张使用，办公会变得更为有效，成本也会降低。20 世纪 80 年代，办公用纸曾一度下降，当时许多需要存档的文书工作被大量转移到计算机上。但随后在 2007 年，办公用纸上升到历史新高，增加的主要部分是复印。2007 年以后，纸张使用逐步减少，这主要得益于人们使用越来越多的移动智能设备和诸如电子签名之类的工具。

尽管早期数字时代人们致力于无纸化办公的乐观愿望还未实现，但电子邮件、文字处理以及电子表格已经让办公环境发生了革命性变化。然而，让电子商务变得切实可行的，还是要归功于互联网的普遍使用。

家喻户晓的例子也许要算网购了。作为顾客，我们享受着在家购物的便利，不用再费时去排队。网购对顾客的不利方面很少，但由于交易类型的不同，与店员之间缺乏沟通可能会抑制在线购物。通过"即时聊天"，在线评论和星级评定，大量的商品和服务选择，以及慷慨的退货政策等在线客户引导工具，与店员缺乏沟通导致的问题正逐步消解。现在，除了购物，我们还能在线支付账单、处理银行业务、购买机票以及在线使用许多其他服务。

易贝的营运方式有些不同，并且值得一提，因为它生成的数据量很大。通过买卖竞价进行交易，易贝每天产生大约50 Tb数据。这些数据是由190个国家或地区的1.6亿活跃用户在其网站上进行的搜索、售卖和竞拍所生成的。通过使用这些数据和适当的分析，易贝现在已经实现了类似于奈飞的推荐系统，本章稍后将进行讨论。

社交网站可为企业提供从酒店和度假到衣物、电脑和酸奶等所有方面的即时反馈。运用这些信息，商家能够明白什么可行，可行程度如何，什么会遭到投诉，以便在情况失控之前解决问题。更有价值的是，根据用户过往的购物及其在网站内的行为预测客户将要购买什么。社交网站，如脸书和推特，收集了大量的非结构化数据，若加以恰当分析，商家也可获得商业利益。

77

猫头鹰（TripAdvisor）等旅游网站也与第三方共享信息。

点击付费广告

现在，专业人士越来越认识到，恰当运用大数据能够获得有用信息并吸引顾客，这些可以通过改进商品促销方式和使用针对性更强的广告来实现。我们只要上网，就不可避免地看到在线广告。我们甚至还可以在诸如易贝等各种竞拍网站上自己免费张贴广告。

点击付费模式，是最受欢迎的广告类型之一。它是一种在你进行在线搜索时弹出相关广告的系统。如果商家想要让其广告随同特定的搜索项显示，他们会向服务提供商出价购买与该搜索词相关联的关键字。商家还会设定每天预算的上限。系统多半会参照商家的出价高低来确定广告的显示顺序。

如果你点击其广告，广告商就必须按照报价向服务提供商支付报酬。商家仅在利益相关方点击其广告时才付费，因此广告必须与搜索项匹配才能吸引网络浏览者点击它们。先进的算法可确保为服务提供商（如谷歌或雅虎）带来最大的收益。实施点击付费广告最著名的，要数谷歌的"关键词广告"。当我们用谷歌搜索时，屏幕侧面自动出现的广告就是"关键词广告"工具生成的。点击付费广告的缺点是，可能会非常费钱，此外，为了让你的广告不占用太多空间，对使用的字符数也有限制。

电子欺诈也是一个问题。例如，竞争对手的公司可能会反复点击你的广告，以耗尽你的每日预算。或者通过使用一种被称为"点击机器人"（clickbot）的恶意计算机程序来生成点击。这种欺诈的受害者是广告商，因为服务提供商的费用照付，而用

78

户并没有参与。不过，由于确保安全性并保证商家有钱可赚最符合提供商的利益，因此服务提供商正在进行大量研究工作以打击欺诈。最简便的方法，大概是跟踪促成一笔买卖需要的点击量。如果点击量突然飙升，或者点击量巨大而没有实际购物，那就有可能存在欺诈性点击。

与这种点击付费的做法不同，定向广告明确基于每个人的在线活动记录。要搞清它是如何运作的，我们先来认真了解一下我在第一章中简要提到的"网络饼干"。

网络饼干

该术语最早出现在1979年，当时操作系统尤尼克斯（UNIX）运行了一款叫作"幸运饼干"（Fortune Cookie）的程序。该程序向基于大型数据库而生成的用户提供随机报价。"网络饼干"有几种形式，所有形式都源自外部，并用于记录网站和/或计算机上的某些活动。当你访问网站时，网络服务器会将一条由存储在计算机中的小文件组成的消息发送到浏览器。此消息就是"网络饼干"的一种，但是还有许多其他种类，例如用作用户认证目的和第三方跟踪的"网络饼干"。

79

定向广告

你在互联网上的每一次点击都会被收集并用于定向广告。

用户数据将被发送到第三方广告网络，并以"网络饼干"的形式存储在你的计算机上。当你单击此网络支持的其他站点时，你以前查看过的产品的广告将显示在屏幕上。使用"光束"（一款火狐的免费附件），你可以跟踪哪些公司在收集你的互联网活动数据。

推荐系统

推荐系统提供过滤机制，基于用户兴趣向他们提供信息。其他类型的推荐系统（不基于用户的兴趣）实时呈现用户都在关注什么，并且通常都以"趋势"的形式来加以显示。奈飞、亚马逊和脸书都使用推荐系统。

向顾客推荐产品的一个流行方式是**协同过滤**。笼统地说，该算法使用收集自单个顾客以往购物和搜索的数据，并将其与其他顾客喜恶商品的大数据库进行比较，以便为进一步的购物提供适当的推荐。不过，简单的比较通常并不能产生良好的效果。试看下面的例子。

假设网上书店向顾客出售烹饪书。推荐所有的烹饪书籍会很容易，但这不太可能确保顾客会购买。书籍太多，顾客无法根据自己的喜好选择。需要一种方法将推荐量减少到顾客可能会实际购买的数量。我们来看看三位顾客，史密斯、琼斯和布朗以及他们的购书情况（图19）。

	《每日沙拉》	《今日意大利面》	《明日甜点》	《大众葡萄酒》
史密斯	购买		购买	
琼　斯	购买			购买
布　朗		购买	购买	购买

图19　史密斯、琼斯和布朗购买的书

推荐系统面临的问题是，应该分别向史密斯和琼斯推荐什么书。我们想知道，史密斯更可能会购买《今日意大利面》还是《大众葡萄酒》。

要做到这一点，我们需要运用通常用来比较有限样本集的统计学方法，即所谓的**杰卡德系数**。它指的是两个集合的交集数与两个集合的并集数的比值。通过交集的占比，该系数可衡量两个样本集之间的相似性。杰卡德距离用于衡量两个集合之间的差异，计算方法为"1减去杰卡德系数"。

我们再来看图19，可以看出史密斯和琼斯买的书有一本是相同的，即《每日沙拉》。他们总共买了三种不同的书：《每日沙拉》、《明日甜点》和《大众葡萄酒》。这样，他们的杰卡德系数为1/3，杰卡德距离为2/3。图20显示了所有可能的客户对的计算结果。

史密斯和琼斯之间的杰卡德系数或相似分数，比史密斯和布朗之间的更高。这表明史密斯和琼斯的购物习惯更为接近——因此我们向史密斯推荐《大众葡萄酒》。我们该向琼斯推荐什么呢？史密斯和琼斯之间的杰卡德系数比琼斯与布朗之间的更高，于是我们向琼斯推荐《明日甜点》。

	相同商品数	购买的不同商品总数	杰卡德系数	杰卡德距离
史密斯和琼斯	1	3	1/3	2/3
史密斯和布朗	1	4	1/4	3/4
琼斯和布朗	1	4	1/4	3/4

图20　杰卡德系数和距离

现在假设客户使用五星级系统对购买进行评分。为了利用这些信息，我们需要找到对特定书籍给予相同评级的其他客户，看看他们还购买了什么以及购买历史如何。每次购物的星级评定如图21所示。

	《每日沙拉》	《今日意大利面》	《明日甜点》	《大众葡萄酒》
史密斯	5		3	
琼 斯	2			5
布 朗		1	4	3

图21 购物星级评定

在该示例中，我们使用了一种被称为**余弦相似度度量**的不同计算方法，该方法也将星级评定考虑在内。对于此计算来说，星级评定表给出的信息代表向量。向量的长度或大小归化为1，不再参与计算。向量的方向用作发现两个向量的相似程度以及谁的星级评定最高。根据向量空间的理论，找到两个向量之间的余弦相似度值。这种计算方法与我们熟悉的三角函数方法大不相同，但基本属性仍然保持余弦取0到1之间的值。例如，如果我们发现两个分别代表星级评定的向量之间的余弦相似度为1，那么它们之间的角就是0，因为余弦（0）=1。这种情况下，它们一定重合，于是可以得出如下结论：他们的趣味相同。余弦相似度数值越大，趣味的相似度也越高。

如果你想看数学细节，本通识读本末尾的"进一步阅读"部分提供有参考书目。在我们看来，有趣的是史密斯和琼斯之间的余弦相似度是0.350，史密斯和布朗之间是0.404。这是先前结果的逆转，先前的结果显示，史密斯和布朗的趣味比史密斯和琼斯的更接近。此矛盾可初步解释为，史密斯和布朗对《明日甜点》的看法,比史密斯和琼斯对《每日沙拉》的看法更接近。

我们将在下一节中介绍奈飞和亚马逊都在使用的协同过滤算法。

亚马逊

1994年,杰夫·贝佐斯创立了卡达布拉网站,但不久后更名为亚马逊。1995年亚马逊网站上线,最初是一家线上书店,现在已发展成为一家在全球拥有3.04亿客户的国际电子商务公司。它生产和销售范围广泛,从电子设备到书籍应有尽有,甚至通过亚马逊生鲜服务提供新鲜食品,诸如酸奶、牛奶、鸡蛋等。它还是一家领先的大数据公司,亚马逊网络服务运用基于海杜普的开发成果,能为企业提供基于云的大数据解决方案。

亚马逊收集的数据包括:买了哪些书?哪些书顾客看了但没有买?他们找书花了多长时间?某本书他们看了多长时间?以及他们保存到购物车里的书是否被最终购买?他们能从这些数据中计算顾客每月或每年购买书籍的花销,还可以确定他们是否为老主顾。早期,亚马逊对收集的数据使用标准统计技术来进行分析。抽取个人样本,基于发现的相似度,亚马逊会向顾客提供更多类似的书籍。2001年,亚马逊研究人员又前进了一步,他们申请了一项名为"项对项协同过滤"技术的专利,并获得了成功。此方法查找相似的商品,而不是相似的顾客。

亚马逊收集大量的数据,包括地址、支付信息,以及个人在亚马逊上浏览过或买过的所有物品的详细信息。亚马逊运用其数据,竭力进行着客户市场研究,以激励顾客把更多的钱花在它那里。例如,就书籍而言,亚马逊不仅提供大量选择,还对单个顾客提出重点建议。如果你订阅了"亚马逊金牌服务"（Amazon Prime），它还会跟踪你观看电影和阅读书籍的习惯。许多客户使用具有GPS功能的智能手机，从而方便了亚马逊收集时间和位置的数据。如上大量数据被用来构建客户画像，从而实现为相似的个人推荐类似的物品。

2013年起，亚马逊开始向广告商售卖客户元数据，以提升其网络服务运营，结果大获成功。对作为云计算平台的亚马逊网络服务来说，安全是至关重要和全方位的。为确保只有得到授权的人才能获得客户账户，亚马逊使用了众多安全技术，比如口令、密钥对和数字签名等。

亚马逊自己的数据同样使用AES（高级加密标准）算法进行多重保护和加密，并存储于世界各地的专用数据中心。运用工业标准SSL（安全套接层），在两台机器——如你的家用计算机和亚马逊——之间建立安全连接。

基于大数据分析，亚马逊在**预期出货**方面独领风骚。其理念是运用大数据预估客户会订购何物。起初，这一想法是为了在订单实际兑现之前将物品运至配送中心。服务稍作延展，物品可随获得免费惊喜礼包的幸运客户订单一并发送。根据亚马逊的退货政策，这不是一个坏主意。可以预见，大多数客户将保留他们订购的商品，因为这些物品是基于其个人喜好，通过使用大数据分析找到的。亚马逊2014年在预期出货方面的专利项目还表明，赠送促销礼品可以买来诚意。为了诚意，也为了通过目标营销增加销售量和缩短交货时间，这一切都让亚马逊相信这种冒险很值得。亚马逊还为无人机送货申请了一项专利，称为"金牌空运"（Prime Air）。2016年9月，美国联邦航空管理局放宽了商业机构放飞无人机的规定，允许他们在高度受控的情况下，飞越操作人员的视野范围之外。这可能是亚马逊寻求在订单提交后三十分钟内发货的第一块敲门砖。也许在你的智能冰箱传感器显示牛奶快要用完之后，无人机配送就开始了。

位于西雅图的"亚马逊购"（Amazon Go）是一家食品便利店，也是第一家此类商店，在此购物无须结账。截至2016年12月，商店只服务于亚马逊员工，原计划2017年1月向公众开放，但已经宣告延期。目前，仅有的技术细节只能通过两年前提交的专利申请获得。专利描述了一种无须逐项结账的系统，在客户购物过程中，其实际购物车的商品详情会被自动添加到虚拟购物车里。只要他们拥有亚马逊账号和安装有"亚马逊购"应用程序的智能手机，他们离店通过过渡区时，电子支付就会自动完成。该系统基于一系列数量众多的传感器，它们用来鉴别物品何时从货架上被拿走，或者何时又放回到了货架上。

该系统会给亚马逊生成大量有价值的商业信息。显然，既然从进店到离店期间的每一步购物行动都被记录在案，亚马逊就能够运用这一数据为你下一次前来做好推荐，方式与其在线推荐系统相仿。不过，考虑到对隐私的保护，有些做法很可能会涉及对隐私的侵犯，比如专利申请中提到的使用人脸识别系统鉴别顾客的手段。

奈飞公司

硅谷的另一家公司——奈飞公司，成立于1997年，起初只是免费邮寄DVD的租赁公司。你选择一张DVD后，可以将另一张添加到队列中，它们会被依次发送给你。你还可以对队列进行优先级排序，这一点很有用。这种服务目前仍然存在且获益颇丰，尽管似乎有逐渐消亡的趋势。如今，奈飞已成为国际互联网流媒体供应商，在190个国家或地区拥有约7 500万订户。2005年，奈飞公司成功扩展，开始提供自己的原创节目。

　　奈飞公司收集和使用大量数据以改进客服，例如在努力提供可靠的电影流媒体的同时，积极向单个客户提供个性化的推荐。推荐是奈飞商务模式的核心，其大部分业务都归功于基于数据分析后为客户所做的推荐。奈飞现在可以跟踪你观看的、浏览的或搜索的内容，以及执行这些操作的日期和具体时间。它还会记录你是否正在使用苹果平板电脑、电视或其他设备。

　　2006年，奈飞公司启动了一项旨在改进其推荐系统的大奖赛。它提供100万美元的奖金，征求能将用户电影评级的预测精度提高10%的协同过滤算法。奈飞提供了超过1亿项的训练数据，用于此次机器学习和数据挖掘竞赛——几乎是可以获得的全部数据。奈飞提供的价值5万美元的中期奖金（进展奖），于2007年由科贝尔团队获得。该团队解决了一个相关的但比较容易的问题。"比较容易"在此是个相对的说法，因为他们的解决方案综合了107种不同的算法后才最终得出两种算法。这两种算法仍在不断改进中，奈飞公司也一直在使用。据测算，这些算法符合1亿条用户电影评分，但要获得全部奖金，算法必须符合50亿条评分。全奖最终于2009年颁给了BPC团队，该团队的算法比现有算法提高了10.6%的预测精度。奈飞公司从未完全采用获奖的算法，主要因为此时他们的商业模式已经发生了改变，变成了我们现在所熟悉的流媒体。

　　在奈飞公司将其商业模式从邮寄服务扩展到通过流媒体提

供电影后，他们便能够收集有关客户喜好和观看习惯的更多信息，而这又能让他们提供更好的推荐。然而，奈飞公司也有背离数字形态的做法。他们在全球雇用了总共大约40位兼职标记员，让他们观看电影，根据内容贴上标签，比如"科幻"或"喜剧"等。这就是电影的分类方式——最初使用人工判断，而不是计算机算法。计算机算法是下一步的事。

奈飞公司运用多种推荐算法，合成了一个推荐系统。所有这些算法都基于公司收集的聚合大数据来进行运算。例如，基于内容的过滤通过分析标记员提供的数据，并根据题材和演员等条件找出类似的电影和电视节目。协同过滤监测你的观看和搜索习惯之类的事情。推荐是基于具有相似口味的观看者所观看的内容，但当有不止一人使用同一个账号时，推荐成功率会下降，常见的情况是家庭的几个成员，他们的口味和观看习惯不可避免地会有所不同。为了解决这一问题，奈飞公司为每个账号创建了多重配置的选项。

互联网点播电视，是奈飞公司的另一个增长点。随着大数据分析的不断发展，它将变得越来越重要。除了收集搜索数据和星级评定之外，奈飞现在还能够记录用户暂停或快进的频率，以及他们是否看完了他们打开的节目。奈飞还监视观看节目的方式、时间和地点，以及其他许多变量，这里无法一一提及。据我们所知，运用大数据分析，奈飞现在甚至能够十分准确地预测客户是否会取消订购。

数据科学

"数据科学家"是送给在大数据领域工作人员的通用头衔。

2012年的《麦肯锡报告》强调了数据科学家的缺乏,估计到2018年,数据科学家的短缺仅在美国就达到19万之多。这种趋势在全世界都很明显,尽管政府在积极推动数据科学技能的训练,但专业知识供需的鸿沟似乎仍在扩大。数据科学正成为大学里热门的学习对象,但是到目前为止,毕业生一直无法满足工商业界的需求,只有工作经验丰富的申请人才可以获得高薪。大数据对商业企业来说事关利润。如果经验不足的数据分析师不堪重负,未能提供预期的积极成果,那么希望很快就会破灭。很多时候,公司都在寻找"万能的"数据科学家,期望他能够胜任从统计分析到数据存储和数据安全的所有工作。

数据安全对任何公司都事关大局,大数据会产生自己的安全性问题。2016年,由于数据安全问题,奈飞取消了第二阶段算法大赛。近年来的其他黑客事件包括,2013年的奥多比公司(Adobe),2014年的易贝和摩根大通银行,2015年的安森保险(一家美国健康保险公司)和卡丰手机商贸,2016年的聚友网(MySpace),还有2012年就遭到黑客侵入,但直到2016年才发现被攻击的全球最大职业社交网站领英(LinkedIn),等等。这仅是少数几例,还有很多公司遭到黑客攻击或者其他类型的安全破坏,导致敏感信息未经授权就被传播。在第七章中,我们将深入探讨一些大数据安全漏洞。

第七章

大数据安全与斯诺登事件

2009年7月，当奥威尔的小说《1984》从亚马逊公司的Kindle上被删除时，Kindle的客户发现，原来艺术作品中的情节真的能够变成现实。在小说《1984》中，一种被称为"记忆洞"的设施专门用来焚毁那些被认为具有颠覆性或者不再需要的文件。文件永久消失，历史被重新改写。亚马逊Kindle事件原本就像一个闹剧，亚马逊和出版商之间的分歧导致了奥威尔的小说《1984》和《动物庄园》被删除，客户成了受害者。客户很不满，他们已经为电子书付了费，并认为电子书归他们所有。由一位中学生和另外一个人提起的诉讼最终得以庭外和解。在和解协议中，亚马逊表示，除了像"司法或监管命令要求删除或修改"这种特殊情况之外，公司将不再从人们的Kindle上删除书籍。亚马逊答应给客户提供退款、发放礼品券，或者恢复被删除的书籍。Kindle电子书不仅无法售卖或出借，而且实际上我们似乎并不能真正拥有。

尽管Kindle事件是对法律问题的一个回应，并非恶意为之，

却表明删除电子文件是何等容易。如果没有印刷件的话，完全
删除任何被看作不想要的或者危险的文本真是易如反掌。如果
你明天拿起一本书的纸质版本，你绝对可以肯定你读到的和今
天的完全一样。但你今天在网上读到的东西，你不能肯定明天
读到时是否依然相同。网上没有绝对的确定性。由于电子文档
很容易被操控，可以在作者不知晓的情况下被修改和更新。在
许多情况下，篡改数据极具破坏力，比如对电子医疗记录的篡
改。甚至设计用来证明电子文件真实性的电子签名，也会遭到
黑客攻击。这些凸显了大数据系统面临的诸多问题，例如如何
确保它们按预期工作，崩溃时可以修复，防止被篡改，以及只有
获得授权的人才可访问等。

　　确保网络及其持有的数据安全，是问题的关键。保护网络
免受未经授权的侵入所采取的基本措施，是安装防火墙，即将网
络与未经授权的通过互联网的外部访问隔离开来。即使网络不
会受到直接攻击，比如病毒和木马的攻击，存储于其中的数据仍
然会有危险，尤其是未加密的数据。一种称为"网络钓鱼"的技
术，通过将恶意代码植入受害人的电脑系统来盗取信息。它的
惯用伎俩是伪装成电子邮件，在邮件中携带可执行文件，或窥探
诸如密码等个人安全信息的插件。总的来说，大数据面临的主
要问题是黑客问题。

　　零售商塔吉特2013年遭到黑客攻击，导致估计1.1亿客户
的资料被盗，包括4 000万人的信用卡明细。据报道，到那一年
的11月底，侵入者已经成功地将恶意软件推送到大部分塔吉特
销售点的机器中，并能够收集用户卡实时交易记录。此时，塔吉
特的安全系统由工作地点在班加罗尔的专家小组每天二十四小

时监控。可疑活动被标记出来，小组联系了位于明尼阿波利斯的一级安全团队。遗憾的是，他们没有及时采取行动。我们下文谈到的家得宝黑客攻击事件使用的技术与此相似，但规模更大，导致了大量数据被盗。

家得宝黑客事件

2014年9月8日，自诩为世界最大家居装修用品零售商的家得宝公司在新闻公报中宣布，其支付系统遭到黑客攻击。在2014年9月18日的更新中，家得宝报告说，攻击使大约5 600万张借记卡或信用卡受到影响。换言之，5 600万张借记卡或信用卡的详细资料被盗走。此案中，黑客首先盗取了商家日志，这让他们轻易就能访问系统——但还只是系统中单个商家的那一部分。网络钓鱼技术助黑客达此目的。

下一步，黑客还需要攻破扩展系统。那时，家得宝使用的是微软的XP操作系统，该系统存在可被黑客利用的固有漏洞。自助付费零售结账系统成了目标，因为在整个系统中该分系统清晰可辨。最后，黑客用恶意软件让7 500个自助结账系统终端感染了病毒，从而盗取客户信息。他们使用一种叫"黑POS机"的特定恶意软件，从感染病毒的终端窃取信用卡或借记卡信息。为安全起见，在销售点终端刷卡时，支付卡信息本该加密。但是很显然，这种点对点加密并未实施，结果导致详细资料对黑客敞开了大门。

当银行侦测到近期在家得宝上的账户欺诈行为时，盗窃事件被发现。银行卡详细资料通过暗网上的网络犯罪专营店被卖出。有趣的是，使用收银机（收银机也刷卡）的人未受到此次攻击的影响。其原因似乎是，在计算机主机中，收银机只有编号，罪犯们还没有将其识别为结账点。假如家得宝自助结账终端也使用了简单编号，这次黑客攻击就有可能避免。话虽如此，但在那个时候，"黑POS机"被视为最高水平的恶意软件，几乎无法被发现。所以一旦黑客获得公开访问系统的机会，最终几乎肯定会将"黑POS机"这种恶意软件成功植入。

史上最大数据黑客事件

2016年12月，雅虎宣布，2013年8月发生了数据泄露，涉及的用户数超过10亿。此事件被称为有史以来最大的个人数据网络盗窃，或至少是有史以来最大的公司数据被盗案。窃贼显然使用了伪造的"网络饼干"，让他们无需密码就能访问账号。此前的2014年还曝光了一次针对雅虎的攻击，那次攻击造成5亿账户被破坏。令人毛骨悚然的是，雅虎声称2014年的黑客攻击是由不知名的"国家资助的行为体"干的。

云安全

大数据安全漏洞几乎每天都在增加。数据盗窃、数据勒索和数据破坏，在这个以数据为中心的世界成为重大关切。关于个人数据安全和所有权有很多细思极恐之处。在数字时代来临之前，我们常常将照片保存在影集中，底片是我们的备份。数字时代来临之后，我们将照片用电子方法存储在我们的电脑硬盘里。存储在硬盘上的照片可能会丢失，我们最好有备份，但起码的安全措施是要让这些文件不能被公开访问。我们很多人现在将数据存储在云端。照片、录像、自制电影都需要很大存储空间，所以从这个角度看，云是有意义的。你将文件存储到云端，⁹³就等于将数据上传到数据中心——更加可能的情况是，数据将被存储在多个中心——不止一个备份被保存下来。

如果你将全部照片存储在云端，以今天系统之复杂，你不太可能会丢失它们。但另一方面，如果你想删除某些内容，照片或是录像，要确保所有备份都被删除也变得困难重重。一般情况下，不依赖提供商你无法做到这一点。另一个重要问题是，设定谁有权访问你上传到云端的数据。如果你想让大数据安全，加密是不可或缺的。

加　密

我们在第五章中简略提到过，加密是指"打乱排列顺序而使文件难以识读"的方法。其基本技术至少可回溯远至罗马时代。苏埃托尼乌斯在他的《罗马十二帝王传》中描述了恺撒大帝是如何使用左移三个字母的方法为文件加密的。使用这种方法，单词"secret"就被加密成为"pbzobq"。这被称为"恺撒加密"，它并不难破解。但即使当今使用的最安全的密码，也将移位用作其算法的一部分。

1997年，当时可公开获得的最佳加密方法——数据加密标准算法（DES），被证明是可以攻破的。这很大程度上是由于计算能力的长足进步和相对较短的56位密钥长度。尽管此加密算法可以提供多达2^{56}种不同的密钥，但还是有可能通过逐一测试直到找出正确密钥，并将信息解密。这种情况1998年就发生了，为了达到此目的，电子前线基金会专门建造了计算机"狂暴破解"，它仅用了差不多二十二个小时就完成了任务。

1997年，美国国家标准和技术协会（NIST）担心，DES缺乏保护最高机密文件所要求的安全性，于是发起了一场公开的、全球性的竞赛，以期找到比DES更好的加密方法。竞赛于2001年结束，高级加密标准算法（AES）脱颖而出。提交的算法名称为"Rijndael"，它是两位比利时原创者琼·戴门和文森特·里杰门名字的合成词。

AES是一种用来给文本加密的软件演算法，密钥长度可选128位、192位或256位。密钥长度为128位时，该算法需要九轮处理，每轮由四个步骤组成，再加上只有三个步骤的最后一轮。AES加密算法是迭代算法，对矩阵执行大量计算——这种计算正好是计算机最为擅长的。不过，我们可以不涉及数学转换来非正式描述一下这一过程。

AES算法首先给我们想要加密的文本加上密钥。此时，我们已不能识读文本，但如果有了密钥，也可以轻易地解码。所以，需要更多的加密步骤以确保安全。下一步运用一种叫作"Rijndael S-盒"的查阅表，将所有字母用另外一个字母来替换。同样，如果我们有了"Rijndael S-盒"，我们也可以反向操作，将文本解密。使用恺撒加密将字母左移，加上最后的排列组合算是完成一轮。然后，用该轮排列组合结果开始另一轮编排，每轮编排使用不同密钥，直到所有轮次结束。当然，我们必须能够解码。对AES算法来说，步骤是可逆的。

密钥长度为192位时，总共需要十二轮加密。为了更加安全，可以使用更长的密钥，比如AES算法的256位密钥。但大多数用户，包括谷歌和亚马逊，觉得AES算法的128位密钥已足够满足大数据安全的需要。AES是安全的，至今尚未被攻破过，导致一些政府部门被迫请求苹果和谷歌这样的大公司提供进入加密材料的后门。

电子邮件安全

据估计，2015年每天发送的电子邮件数超过2 000亿，除去垃圾邮件或恶意企图邮件，真正意义上的邮件不足10%。大部分电子邮件是不加密的，其内容易于受到黑客拦截。假如我从加利福尼亚发送非加密邮件到英国，邮件被分成数据"包"，通过邮箱服务器传输，而邮箱服务器是连接到互联网的。互联网其实是由一张庞大的全球电线网络组成，地上的、地下的以及海底的，再加上手机发射塔和卫星。唯一未被连接跨洋电缆的大陆是南极洲。

因此，尽管互联网和云计算通常被认为是无线的，其实根本不是。数据是通过铺设在海底的光纤电缆传输的。各大洲之间的数字通信几乎全都如此。即使使用云计算服务，我的邮件还是要通过跨大西洋光纤电缆传输。云技术，这个诱人的时髦词，让人联想起的图景是通过卫星向全世界发送数据，但实际上，云服务是牢牢扎根于主要通过电缆提供互联网访问的分布式数据中心网络。

光纤电缆是数据传输的最快方式，因而通常比卫星更可取。当前对光纤技术的大量研究让数据传输速度越来越快。跨大西洋电缆已经成为一些好奇和意外攻击的对象，包括想咬断电缆的鲨鱼的攻击。按照国际电缆保护委员会的说法，鲨鱼攻击在有记录的故障中占比不到1%，但即便如此，易受攻击地区的电缆现在常常用凯夫拉纤维加以保护。假如跨大西洋电缆没有因为好奇的鲨鱼、敌意政府或粗心渔民发生故障，我的电子邮件将登陆英国并继续前行，那么也许就在这个节点上会像其他互联网数据一样被截获。2013年6月，爱德华·斯诺登泄露的文件表明，英国政府通信总部（GCHQ）挖掘了大量数据，这些数据是利用一种叫作"Tempora"的系统通过大约200条跨大西洋电缆收到的。

斯诺登事件

爱德华·斯诺登是美国计算机专家，2013年因泄露美国国家安全局的机密信息而被控犯有间谍罪。这起备受瞩目的案件引发了公众对政府大规模监听能力的关注，同时也引起了人们对个人隐私的普遍担忧。斯诺登自泄密以来广获殊荣，他曾当选格拉斯哥大学校长，被评为《卫报》2013年年度人物，并获得了2014年、2015年和2016年的诺贝尔和平奖提名。大赦国际支持斯诺登，认为他是一个为国履职的揭发者。但是，美国政府官员和政界人士对此却并不苟同。

2013年6月，英国《卫报》报道称，美国国家安全局一直从美国一些重要的电话网络中收集元数据。很快，一个名为"棱镜"的项目被曝光，该项目针对的是与美国通信的外国公民，旨在收集并存储他们的互联网数据。之后，一系列电子泄密事件被公之于众，矛头直指美国和英国政府。爱德华·斯诺登正是披露这些信息的人，他当时是博思艾伦咨询公司的雇员，也是在夏威夷密码中心工作的美国国家安全局承包商。斯诺登把这些信息发给了他认为值得信任的媒体人，以避免在未经审慎思考的情况下贸然公布。斯诺登的动机和其中牵涉的法律问题并非本书的讨论范围，但很明显，斯诺登认为，美国最初对其他国家的合法监视现已转变为监视美国自己了，美国国家安全局现在正对所有美国公民进行非法监视。

免费的万维网刮网工具"悉数下载"是火狐浏览器的一个可用扩展程序，它与"wget"程序都为用户提供了快速下载网站全部内容或其他万维网相关数据的便利。作为美国国家安全局机密网络的授权用户，斯诺登利用这些现有的应用程序下载并复制了大量信息。他还把数量庞大且极其敏感的数据从一个计算机系统转移到了另一个计算机系统。为完成上述工作，需要用户名和密码，作为系统管理员，他拥有这些敏感信息。因此，他能轻易接触到许多机密文件并进行窃取，除了密级更高的文件之外。为了获得比绝密文件等级更高的文件，他必须使用更高级别用户账户的身份验证信息，安全协议本应该阻止斯诺登获取这些信息。然而，由于是斯诺登本人创建了这些账户并拥有系统管理员的特权，所以他知道详细信息。斯诺登还成功说服了至少一名安全级别高于他的美国国家安全局雇员，并成功获知了他们的密码。

最终，斯诺登复制了约150万份高机密文件，并将其中约20万份文件交给了值得信任的记者，尽管最终只有很少的文件被公布。（斯诺登明白，他盗取的文件并不都能公之于众，哪些能公布，他也很谨慎。）

虽然斯诺登从未完全披露具体细节，但他似乎能将数据复制到闪存盘上，显然，他下班时随身携带闪存盘也没什么困难。安全措施明显不足以防止斯诺登复制这些文件。即便是在离开 98 时对他进行一次简单的全身扫描，也能侦测出他是否携带有任何便携式设备，而且办公室的视频监控也应该能记录下可疑活动。2016年12月，美国众议院解密了一份2016年9月的文件，该文件虽有大量的修改痕迹，却也评估了斯诺登其人以及被泄文件的性质和影响。我们从这份文件中可以清晰地看出，美国国家安全局并未采取足够的安全措施，此后，"维护网络安全倡议"开始实施，虽然离全面施行还有待时日。

斯诺登拥有许多系统管理员特权，但因这些数据极为敏感，允许一个人不受安保约束的随意访问是不可接受的。假如要访问或传输数据时需要两个人的验证凭据，这或许就足以预防斯诺登非法复制文件。另外，斯诺登显然能够插入优盘，复制任何他想要的东西，这也很奇怪。对此，一个非常简单的安全措施就是禁用DVD和USB端口，或是在最开始就不进行安装。此外，若将视网膜扫描作为进一步的身份验证以优化密码安全性，斯诺登就很难访问那些高密级文件。现代安全技术非常先进，若使用得当，系统便很难被渗透。

2016年末，在谷歌搜索中输入"爱德华·斯诺登"，短短一秒钟就有2 700多万条搜索结果；搜索词"斯诺登"有4 500万条搜索结果。由于许多网站允许访问被泄文件，甚至公布了标有"绝密"字样的泄露文件，它们现在稳稳地处于全球公共领域之内，并毫无疑问还将继续如此。爱德华·斯诺登目前居住在俄罗斯。

与斯诺登事件相比，维基解密的故事就非常不同了。

维基解密

维基解密是一个庞大的在线揭秘组织，旨在传播秘密文件。

其运转资金来自捐赠，员工主要由志愿者组成，不过似乎也曾雇用过一些员工。截至2015年12月，维基解密声称已发布（或泄露）了1 000多万份文件。维基解密通过自己的网站、推特和脸书维持着极高的知名度。

2010年10月22日，维基解密公布了大量代号为"伊拉克战争日志"的机密数据（共391 832份文件），维基解密及其创始人朱利安·阿桑奇也因此登上了新闻头条，引起了极大争议。此前，2010年7月25日，就有约7.5万份名为"阿富汗战争日记"的文件被泄露。

美国陆军士兵布拉德利·曼宁对这两起泄密事件负有责任。他在伊拉克做情报分析员时，曾随身携带一张光盘，从一台本应安全的个人电脑上复制了机密文件。因此，布拉德利·曼宁，也就是改名后的切尔西·曼宁①，在2013年因违反《反间谍法》和其他相关罪行被军事法庭判处三十五年监禁。2017年1月，美国前总统巴拉克·奥巴马在离任前为切尔西·曼宁减刑。"曼宁女士"在狱中曾因患有性别焦虑症而接受治疗，她于2017年5月17日获释。

维基解密虽然受到了政客和政府的严厉批评，但大赦国际（2009）和英国《经济学人》（2008）等诸多机构都对维基解密进行了赞扬和奖励。这些机构的网站信息显示，朱利安·阿桑奇在2010年至2015年间，连续六年获得诺贝尔和平奖提名。诺贝尔委员会通常在五十年后才会公布被提名者名单，但提名人（自

---

①　曼宁原本是男性，全名为布拉德利·曼宁。在被判入狱后，他表明自己从小就有女性的性别认同，于2013年8月22日发表公开声明，宣布改名为"切尔西·曼宁"，开始荷尔蒙治疗，并转变性别为女性。——译注

身必须达到和平奖委员会的严格标准）往往会公开他们的提名人选。例如，2011年，挪威国会议员斯诺里·瓦伦提名了朱利安·阿桑奇，并对维基解密揭露所谓的侵犯人权行为表示支持。2015年，阿桑奇得到了前英国国会议员乔治·加洛韦的支持。2016年初，一个支持阿桑奇的学术团体也呼吁授予他诺贝尔和平奖。

但是，2016年底出现了反对阿桑奇和维基解密的浪潮，其中部分原因是他们的报告被指存有偏见。对维基解密不利的是道德关切，涉及个人安全和隐私、公司隐私、政府秘密、冲突地区的地方资源保护及一般公共利益等。对朱利安·阿桑奇和维基解密来说，事态正变得越来越混乱。例如，希拉里的电子邮件在2016年被泄露的时间，正是破坏希拉里·克林顿总统候选人资格的最佳时机，这使得人们开始质疑维基解密的客观性，一些广受尊敬的人士也因此对其大加批评。

无论你对阿桑奇和维基解密是支持还是谴责（根据所涉及的具体问题，支持和谴责尽皆有之），还有一个重大的技术问题是，是否有可能关闭维基解密网站。由于维基解密将数据存放在世界各地的诸多服务器上，且其中一些服务器还位于同情维基解密的国家，所以即使大家都希望如此，想要完全关闭维基解密也是不太可能实现的。不过，出于对自身的保护，防止在每次解密后遭到报复，维基解密发布了保险文件。其潜台词是，如果阿桑奇出了什么事，或者维基解密被关闭，那么保险文件的密钥将会被公之于众。维基解密最新的保险文件使用AES加密法，密钥为256位，因此不可能被攻破。

2016年，爱德华·斯诺登与维基解密之间出现了分歧，其

101

争执在于他们泄露数据的不同方式。斯诺登会把文件交给他所信任的记者，这些记者会仔细选择要披露的文件内容。美国政府官员会事先得到消息，揭秘方会按照他们的建议，保留部分文件，以维护国家安全。迄今为止，许多文件还从未被披露过。而维基解密似乎只是简单地公布数据而已，他们几乎没有考虑到要保护个人信息。维基解密仍寻求从告密者处收集信息，但最近披露的数据是否可靠，或所选择披露的信息是否完全公正，都不得而知。维基解密在其网站上教授人们如何使用一种叫作"洋葱路由"（TOR）的工具来匿名发送数据和确保隐私，但使用"洋葱路由"的人并非都是揭秘者。

洋葱路由和暗网

普林斯顿大学社会学系助理教授珍妮特·维尔特斯决定在自己身上做个实验，看看能否对网络营销人员隐瞒自己怀孕的事实，从而防止个人信息进入大数据库。维尔特斯博士曾在2014年5月的《时代》杂志上讲述了她的此次经历。她采取了一些特殊的隐私保护措施，包括不使用社交媒体；下载了洋葱路由，并用它订购了许多婴儿用品；并且在商店购物时都使用现金支付。虽然她所做的一切都是完全合法的，但最终她得出的结论是，选择逃避大数据的追踪费时费力，用她自己的话来说，这让她看起来像是一个"坏公民"。然而，洋葱路由还是值得一用，尤其是因为它让维尔特斯博士感到安全，且有效防止了她的隐私被追踪。

洋葱路由是一个加密的服务器网络，最初由美国海军开发，旨在提供一种匿名使用互联网的手段，从而防止个人数据被追踪和收集。洋葱路由的项目仍在进行,其目的是开发和改善任何关注隐私的人都可以使用的开源在线匿名环境。洋葱路由的工作原理是对你的数据(包括发送地址)进行加密,然后通过删除部分标头(关键是部分IP地址)对数据进行匿名处理,因为利用这些信息进行回溯,很容易就能找到发送者。生成的数据包在到达最终目的地之前,都通过志愿者托管的服务器或中继系统传输。

从积极方面看,洋葱路由的用户包括最初设计它的军方,希望保护信息来源和信息本身的调查性新闻报道记者,以及希望保护个人隐私的普通公民。商家利用洋葱路由守住商业秘密,政府利用它保护敏感信息来源和信息本身。有一份洋葱路由项目的新闻稿列出了1999年至2016年间与洋葱路由相关的一些新闻报道。

从消极方面看,洋葱路由匿名网络已经被网络罪犯广泛使用。用户可以通过洋葱路由提供的服务匿名访问域名后缀为"onion"的网站。这类网站很多都极为糟糕,其中包括用于毒品交易、色情服务和洗钱的非法暗网。例如,广为人知的"丝绸之路"网站就能通过洋葱路由访问,该网站属于暗网且是非法毒品供应商。正因为洋葱路由的使用,执法人员难以进行追踪。罗斯·威廉·乌布利希被捕后,法院审理了这起重大案件。之后,他因以"恐怖海盗罗伯茨"的化名创建并经营"丝绸之路"网站而被定罪。该网站被关闭,但后来又死灰复燃。2016年,该网站以"丝绸之路3.0"的名字迎来了第三次重生。

深　网

深网指的是那些不能被谷歌、必应和雅虎等普通搜索引擎

103

索引到的网站。深网既包括合法网站，也包括暗网网站。人们普遍认为，深网比我们熟悉的表层网络要大得多，即使使用特殊的深网搜索引擎，也很难评估这个隐藏的大数据世界具有何等规模。

104

第七章　大数据安全与斯诺登事件

第八章

大数据与社会

机器人和工作机会

　　1930 年英国经济萧条期间，著名经济学家约翰·梅纳德·凯恩斯著文，推测一个世纪以后的职业生活会是怎样的情形。工业革命创造了以城市为基础的工厂里新的工作机会，并让农耕社会发生了很大的改变。人们认为，劳动密集型工作将最终由机器来完成，一些人会失业，另一些人的工作时间会大大缩短。由于技术进步，人们可以减少有偿劳动而获得更多的闲暇，凯恩斯尤其关注人们将如何利用所获得的闲暇时间。也许更为迫切的是财务支持问题，有人提议实施全民基本工资以应对工作机会减少的窘境。

　　在 20 世纪，我们逐渐看到工厂里的工作机会被越来越精密的机器所侵蚀。尽管很多生产线几十年前就已经自动化了，但凯恩斯主义者每周工作十五小时的理想尚未实现，并且在近期也似乎不太可能实现。正如工业革命一样，数字革命将不可避

免地改变就业状况,但改变的方式我们还不能准确预测。随着
"物联网"技术的进步,我们的世界继续变得越来越受数据驱动。
使用实时大数据分析的结果来指导决策和行动,将在我们的社
会中发挥越来越重要的作用。

有人认为,建造机器和给机器编码还是离不开人。但这也
仅仅是猜测。无论如何,这只是专业工作的一个领域。即使在
这个领域里,我们也可以很现实地预期,机器人会越来越多地
取代人类。比如,复杂的机器人医学诊断会减少医务人员。出
现像"沃森"那样具有人工智能的机器人外科医生,是完全可能
的。自然语言处理(另一个大数据领域)将发展到我们无法分
辨是在与机器人还是在与医生对话——至少在我们不面对面的
时候。

不过,很难预测一旦机器人接管了很多现有的岗位之后,人
类还有哪些工作岗位。创造力被认为属于人类。但计算机科学
家通过与剑桥大学和阿伯里斯特维斯大学合作,研发出了"亚
当",一款机器人科学家。"亚当"已在基因组学领域成功提出和
检验了新的假说,做出了新的科学发现。曼彻斯特大学团队成
功研发出"夏娃",一款用于热带疾病药品设计的机器人,使类
似研究又前进了一步。这两项工程都使用了人工智能技术。

小说家的技巧似乎为人类所独有,它依靠经验、情感和想象
力。但就是这一创造性领域,也正受到机器人的挑战。日经新
闻文学奖接受由非人类作者写作或合写的小说。2016年,四部
由人和计算机联合写就的小说通过了评奖初选,评委对作者身
份并不知情。

尽管科学家和小说家最终可能都要与机器人合作,但对我

们大多数人来说，大数据驱动型环境更显著的影响，是通过智能设备出现在我们的日常生活中。

智能交通工具

2016年12月7日，亚马逊宣布成功实施了首次商业无人机通过GPS送货。收货人是居住在英国剑桥附近乡村的一位男子，他收到了一个重4.7磅的包裹。无人机送货目前仅能提供给两个亚马逊"金牌空运"客户。两家都住在剑桥附近投送服务中心5.2平方英里范围之内。"进一步阅读"部分提到的录像，显示的就是此次无人机飞行。这似乎吹响了该计划进行大数据收集的号角。

亚马逊并不是第一家成功进行商业无人机送货的公司。2016年11月，福乐梯公司就开始了在新西兰大本营小范围的无人机投送比萨饼服务，在其他地方也有类似项目。目前来看，无人机投送服务可能会有所增加，尤其是在安全问题易于管理的偏远地区。当然，网络攻击，或仅仅是计算机系统的一次故障，就会造成大破坏，假如小小的送货无人机运转失灵，就很可能会造成人员或动物伤亡，以及惨重的财产损失。

这种情形与在公路上以时速70英里行驶的汽车被软件远程接管如出一辙。2015年，《连线》杂志的两位安全专家查理·米勒和克里斯·瓦拉塞克使用志愿者进行了一项测试，结果显示，用来连接汽车和互联网的汽车仪表盘电脑"U链接"在汽车行驶时可以被黑客远程控制。这份报告引起轰动，两位黑客专家能够使用笔记本电脑的互联网连接并控制一辆切诺基吉普车的方向、制动和变速器，直至其他非关键功能诸如空调和收音机等。吉普车在繁忙的公路上以每小时70英里的速度行驶，突然间，加速装置完全失效，驾驶员志愿者惊恐万分。

这次测试后，汽车生产商克莱斯勒公司向1 400万辆汽车车主发出了警告，并赠送优盘，内含通过仪表盘端口安装的软件更新。这次攻击是因为智能手机网络存在漏洞。之后，漏洞被修复，但这件事表明，智能车辆技术在完全走近大众之前，遭到网络攻击的潜在风险需要提前得到解决。

从汽车到飞机等自动交通工具的使用似乎已势不可挡。飞机已经能够自动飞行，包括起飞和降落。尽管离用无人机大量运送乘客还差一步，但无人机目前已开始用于农业的智能作物喷施，以及军事用途。智能交通工具的广泛应用仍然处于发展的早期阶段，但智能设备业已成为现代家居的一部分。

智能家居

正如第三章中所提到的"物联网"这一术语，是指连接互联网的大量电子传感器。比如，任何可以在家中安装并远程管理的电子设备——通过人机交互界面显示在住户的电视屏幕、智能手机或者笔记本电脑上——都是智能设备，因而也就是物联网的一部分。很多家居都安装了声控中心控制点，用以管理照明、供暖、车库门以及其他各种家用电器。一旦有了Wi-Fi（"无线保真"的缩写，表示使用无线电波而不是电线连接诸如互联网等网络的能力）链接，你就能让智能音箱（通过叫它的名字来实现，你会给它取个名字的）告诉你当地的天气情况，或者报道国内新闻。

这些设备提供基于云的服务。但说到隐私，它们还是有缺陷的。只要设备电源一打开，你说的任何话都被记录下来并存储到遥远的服务器上。在最近的一次谋杀调查中，美国警方要求亚马逊提供"回声"智能音箱（该设备通过语音控制连接到Alexa语音服务，可播放音乐、提供信息和新闻报道等）数据。亚马逊起初不愿这么做，但嫌疑人最近同意让他们发布录音，希望这些录音能够证明他的清白。

基于云计算的进一步发展，意味着诸如洗衣机、冰箱以及家用清洁机器人等电器，都将成为智能家居的一部分，可通过智能手机、笔记本电脑或家庭扬声器远程遥控。由于所有这些系统都是通过互联网运行，因而也就有黑客入侵的潜在风险。安全也因此成了很大的研究领域。

说到黑客入侵，即便儿童玩具也未能幸免。一款被称为"我的朋友凯拉"的智能玩偶，被伦敦玩具工业协会推举为"2014年度创新玩具"，它就遭遇了黑客入侵。通过隐藏在玩偶中的蓝牙设备，儿童可对玩偶提问并能听到回答，但这个蓝牙设备并不安全。负责监督互联网通信的德国联邦网络管理局敦促家长销毁这种玩偶。目前玩偶已被禁售，因为它威胁到个人隐私。黑客们能够很容易听到孩子的问题并给出不恰当的答案，包括生产厂家列单禁止的那些词汇。

智慧城市

尽管智能家居才刚刚开始走进现实生活，但人们预测物联网加上多种信息和通信技术（ICTs）会将智慧城市变成现实。很多国家，包括印度、爱尔兰、英国、韩国、中国和新加坡都已经开始规划智慧城市。它们的想法是，由于城市在快速扩张，必

109

须要在拥挤的世界建立更加高效的智慧城市。农村人口正以越来越快的速度拥入城市。2014年的时候，大约54%的人口生活在城市。联合国预测，到2050年，世界人口的66%将居住在城市里。

从物联网和大数据管理技术实施的早期开始，就有零星但越来越多的想法推动着智慧城市技术的发展。比如无人驾驶汽车、远程健康监控、智能家居以及电子通勤等都将成为智慧城市的特征。此类城市的运作将依靠对大数据的管理和分析，它们从大量布设的传感器收集而来。大数据加上物联网的助力，是智慧城市的关键。

总的来说，智慧城市的好处之一就是智能能源系统。它能规范街道照明、监控交通状况，甚至跟踪垃圾去向。通过在全城安装大量射频识别标签和无线传感器，这些都可以做到。这些标签由芯片和微型天线组成，将数据从单个设备发送到数据中心进行分析。比如，城管部门通过在车辆上安装射频识别标签和在街道安装数字摄像头来监控交通状况。提高人身安全也是考虑的对象，比如通过悄悄对孩子进行标识，父母即可用智能手机对他们进行监控。这些传感器会产生大量数据，它们需要通过中央数据处理装置进行实时监控和实时分析，然后可用于多种多样的目的，包括测量交通流量、识别拥堵以及推荐备选线路。在这种情况下，数据安全就显得举足轻重，因为系统的任何重大故障或黑客入侵都会很快影响公众信心。

计划于2020年完工的韩国松岛国际商务区，是专门建造的智慧城市。它的主要特征之一是，整座城市拥有光纤宽带。这项先进的技术用来确保快速享有智慧城市的福利。新的智慧城市还致力于将负面环境影响最小化，使其成为未来可持续发展的城市。虽然已经规划了很多智慧城市，并且像松岛那样的城市正在专门建设当中，但现有城市皆须将其基础设施逐渐现代化才能适应时代的要求。

2016年5月，联合国全球脉搏，一项旨在推进让全球受益的大数据研究，在东盟十国及韩国揭开了"2016年创意大赛：可持续城市"竞赛的序幕。到截止时间的6月份，共收到超过250份提案，各个领域的获胜者已于2016年8月公布。韩国因其通过利用关于排队的众包信息，达到减少等待时间，从而改进公共交通的建议斩获头奖。

展　望

从这本"通识读本"，我们看到了由于互联网和数字化世界的发展带来的技术进步，以及由此而来的在过去几十年中数据科学发生的急剧变化。在这最后一章，我们领略了未来的生活可能被大数据重塑的方方面面。尽管我们不能指望这本简短介绍覆盖大数据产生影响的所有领域，但我们还是看到了其中一些已经在影响我们生活的各种不同的应用。

我们生活的世界所生成的数据会越来越大。有效且有意义的处理大数据的方法，毫无疑问将是我们继续深入研究的课题，尤其是在实时分析领域。大数据革命标志着世界的运行方式发生了重要改变。随着所有这些技术的发展，个人、科学家和政府共同担负着确保其恰当使用的道德责任。大数据是力量，它的潜力是巨大的。如何避免其被滥用，取决于我们自己的努力。

字节大小量表

术 语 名 称	含 义
Bit 位或比特	存放一位二进制数，即 0 或 1，最小的存储单位
Byte 字节	8 个二进制位为一个字节，最常用的单位
Kilobyte（Kb）	等于 1024 b
Megabyte（Mb）	等于 1024 Kb
Gigabyte（Gb）	等于 1024 Mb
Terabyte（Tb）	等于 1024 Gb
Petabyte（Pb）	等于 1024 Tb
Exabyte（Eb）	等于 1024 Pb
Zettabyte（Zb）	等于 1024 Eb
Yottabyte（Yb）	等于 1024 Zb

113

小写英文字母 ASCII 码表

十 进 制	二 进 制	十六进制	字 母
97	01100001	61	a
98	01100010	62	b
99	01100011	63	c
100	01100100	64	d
101	01100101	65	e
102	01100110	66	f
103	01100111	67	g
104	01101000	68	h
105	01101001	69	i
106	01101010	6A	j
107	01101011	6B	k
108	01101100	6C	l
109	01101101	6D	m
110	01101110	6E	n
111	01101111	6F	o

十 进 制	二 进 制	十六进制	字 母
112	01110000	70	p
113	01110001	71	q
114	01110010	72	r
115	01110011	73	s
116	01110100	74	t
117	01110101	75	u
118	01110110	76	v
119	01110111	77	w
120	01111000	78	x
121	01111001	79	y
122	01111010	7A	z
32	00010000	20	空格

小写英文字母 ASCII 码表 116

索 引

（条目后的数字为原书页码，
见本书边码）

A

ACID 原子性、一致性、独立性和持久性 29, 34, 36

Advanced Encryption Standard 高级加密标准 (AES) 84, 95, 101

Adwords（谷歌）关键词广告 78, 79

Afghan War Diary, The 阿富汗战争日记 100

Amazon 亚马逊（公司）36, 37, 44, 56, 80—86, 90, 95, 107, 109

Amnesty International 大赦国际 97, 100

Animal Farm《动物庄园》90

American Standard Code for Information Interchange 美国信息交换标准码 (ASCII) 38, 39

Apollo 11 "阿波罗 11" 号 26

Armstrong, Neil 尼尔·阿姆斯特朗 26

Assange, Julian 朱利安·阿桑奇 100, 101

autonomous vehicle 自动驾驶车辆 10, 11, 108

B

Babylonians 巴比伦人 2

BASE 基本可用、软状态和最终一致三要素 34

BellKor Pragmatic Chaos BPC 团队 87

Berners-Lee, T. J. 伯纳斯—李 17

Bezos, Jeff 杰夫·贝佐斯 83

Bing 必应 7, 104

BlackPOS 黑 POS 机（恶意软件）92

blogs 博客 60

Bloom filter 布隆过滤器 47—51

Booz Allen Hamilton 博思艾伦咨询公司 97

Box, George 乔治·博克斯 56

Brewer, Eric 埃里克·布鲁尔 33

Brin, Sergey 谢尔盖·布林 51, 56

Business Week《商务周刊》76

byte 字节 6, 12, 38, 39, 42, 48

C

Caesar cipher 恺撒加密法 94, 95

Caesar, Julius 尤利乌斯·恺撒 94

Cafarella, Mike 迈克·卡弗雷拉 30

CAP Theorem CAP 定理 33—34

census 人口普查 2, 4

CERN 欧洲粒子物理研究所 16

cholera outbreak 霍乱暴发 3

Chrysler 克莱斯勒（公司）108

classification 分类 23—25, 68

clickstream logs 点击流日志 7, 8

Clinton, Hillary 希拉里·克林顿 101

Cloud, The 云 28, 36, 37, 74, 83, 84, 93, 94, 96, 109

cluster 集群 3

clustering 聚类 22, 23, 31

collaborative filtering 协同过滤 80, 83, 84, 87, 88

Common Crawl "爬网侠"（数据平台）46, 56

compression 压缩 37—42

cookies "网络饼干" 8, 79, 93

correlation 相关性 19, 57, 58, 62

credit card fraud 信用卡欺诈 21, 22, 24, 25, 51, 91, 92

Crick, Francis 弗朗西斯·克里克 10

Cutting, Doug 道格·卡廷 30, 31

D

Daemen, Joan 琼·戴门 95

dark Web 暗网 93, 102—104

Data Encryption Standard 数据加密标准算法 (DES) 94, 95

data mining 数据挖掘 20, 23, 70, 87

data science 数据科学 13, 88

data scientist 数据科学家 13, 88, 89

datum 数据 3

Deep Crack 狂暴破解 94

deep Web 深网 104

Delphi Research Group 德尔菲研究小组 65

digital marketing 数字化营销 8, 85

Discrete Cosine Algorithm 离散余弦算法 41

distributed file system 分布式文件系统 30, 44, 45

Down Them All 悉数下载 98

Dread Pirate Roberts 恐怖海盗罗伯茨 103

drones 无人机 85, 108

E

earthquake prediction 地震预测 13

Ebola 埃博拉 45, 46, 65—67

e-commerce 电子商务 76—78

Economist The《经济学人》100

Electronic Delay Storage Automatic Computer 电子延迟存储自动计算器 (EDSAC) 75

Electronic Frontier Foundation 电子前线基金会 94

email 电子邮件 6, 9, 17, 48—51, 73, 77, 91, 92, 96, 97, 101

encryption 加密 73, 74, 84, 91, 92, 94, 95

Espionage Act《反间谍法》100

exabyte 百亿亿字节 (Eb) 6

F

Facebook 脸书（网站）8, 31, 32, 44, 52, 60, 78, 80, 100

FancyBears 奇幻熊（网站）73

fibre-optic cable 光纤 96, 111

firewall 防火墙 91

Fisher, Ronald 罗纳德·费希尔 14, 15, 18

Flirtey Inc. 福乐梯公司 107

G

Gauss, Carl Friedrich 卡尔·弗里德里希·高斯 3

索引

global pulse（联合国）全球脉搏 111

Google 谷歌（公司）7, 30, 44, 51, 52, 56, 58, 60—65, 78, 95, 99, 104

Google Flu Trends 谷歌流感趋势 60—65, 67

Government Communications Headquarters (GCHQ)（英国）政府通信总部 97

Guardian, The《卫报》97

H

hacking 黑客行为 37, 74, 91, 93, 111

Hadoop 海杜普 30—32, 43—45, 83

Hammond, Henry 亨利·哈蒙德 3

Henry Samueli School of Engineering and Applied Science, UCLA 加州大学洛杉矶分校亨利·萨穆利工程与应用科学学院 43

Hollerith, Herman 赫尔曼·何乐礼 4

Home Depot 家得宝（公司）92, 93

Human Genome Project 人类基因组计划 10, 69—70

Humby, Clive 克莱夫·胡姆比 20

I

International Business Machines Corporation 国际商用机器公司 (IBM) 4, 6, 16, 26, 68, 70, 71, 76

International Computers Limited 国际计算机有限公司 (ICL) 76

International Data Corporation 国际数据公司 (IDC) 37

Internet of Things 物联网 28, 108 122

Iraq War Logs 伊拉克战争日志 100

Ishango Bone 伊山戈骨 2

J

Jaccard index 杰卡德系数 81, 82

Java 爪哇（编程语言）31

Jeopardy《危险边缘》（电视节目）70—72

J. Lyons and Co. 里昂公司 75

Joint Photographic Experts Group 联合图像专家小组格式 (JPEG) 42

K

Kaptoxa 黑 POS 机（恶意软件）92, 93

Keynes, John Maynard 约翰·梅纳德·凯恩斯 105

Kindle（亚马逊）电子书阅读器 90

Kuhn, Thomas 托马斯·库恩 57

L

Laney, Doug 道格·莱尼 16, 18

Laplace, Pierre–Simon 皮埃尔—西蒙·拉普拉斯 3

Large Hadron Collider 大型强子对撞机 16, 17

Lavoisier, Antoine 安托万·拉瓦锡 3

Lebombo Bone 列朋波骨 2

lossless compression 无损压缩 37—40

lossy compression 有损压缩 38, 40—43

大数据

M

machine learning 机器学习 20, 71, 87

Magnetic Resonance Imaging 核磁共振 (MRI) 68, 70

magnetic Tape 磁带 75

Manning, Bradley 布拉德利·曼宁 100

Manning, Chelsea 切尔西·曼宁 100

MapReduce 映射归约 , 43—47

marketing 营销 8, 12, 20, 85

market research 市场研究 84

Miller, Charlie 查理·米勒 107

Millionaire calculator "百万富翁" 计算器 15

Moore, Gordon 戈登·摩尔 27

Moore's Law 摩尔定律 27, 28

My Friend Cayla 我的朋友凯拉(智能玩偶) 109

N

National Institute of Standards and Technology (美国) 国家标准和技术协会 (NIST) 94

National Security Agency (美国) 国家安全局 (NSA) 97

Natural language processing 自然语言处理 (NLP) 71

Nepal Earthquake 尼泊尔地震 67—68

Netflix 奈飞(公司) 28, 37, 77, 80, 83, 86—89

New SQL 新结构化查询语言 36

Newton, Isaac 艾萨克·牛顿 3, 57

Ngram N 元(浏览器) 56

Nikkei Hoshi Shinichi Literary Award 日经新闻文学奖 106

1984 《1984》 90

Nissan 日产(公司) 10

Nobel Peace Prize 诺贝尔和平奖 97, 100

normalization 规范化 29

NoSQL 非关系型数据库 33—36

O

Orwell, George 乔治·奥威尔 90

over-fitting 过拟合 63

Oxford English Dictionary 《牛津英语词典》 3

P

Page, Larry 拉里·佩奇 51, 56

PageRank 佩奇排名(网页排名)名算法 51—56

pay-per-click advertising 点击付费广告 78, 79

phishing 网络钓鱼 48, 73, 91, 92

population 人口 2, 3, 8, 15, 66—68, 110

Priestley, Joseph 约瑟夫·普里斯特利 3

PRISM 棱镜 97

public datasets 公共数据集 56

punched cards tabulator 打孔卡制表机 4

Q

Quipu 奇普 2, 3

R

radio-frequency identification 射频识别 (RFID) 110

recommender systems 推荐器系统 77, 80—83, 87, 88

regression 回归 68

relational database 关系型数据库 29, 30, 32, 34, 36

Rijmen, Vincent 文森特·里杰门 95

Rijndael algorithm Rijndael 加密算法 95

Rijndael S-Box Rijndael S-盒 95

robots 机器人 105—107

S

sample data 抽样数据 15, 89

sample survey 抽样调查 4

scalability 可扩展性 29, 33, 36

security 安全性 9, 69, 70, 73, 74, 79, 84, 89—111

semi-structured data 半结构化数据 5, 6, 17, 18, 31, 34

sensor data 传感器数据 17, 18, 59, 68

Silk Road 丝绸之路(网站)103

Snow, John 约翰·斯诺 3, 4

Snowden, Edward 爱德华·斯诺登 90, 97—100, 102

social media 社交媒体 18, 60, 102

social networking 社交网络 6, 8, 17, 35, 59, 60, 77, 78

Songdo International Business District, South Korea 韩国松岛国际商务区 111

Sparta 斯巴达 1

Square Kilometer Array Pathfinder (澳大利亚)平方公里阵列探路者 (ASKAP) 12

structured data 结构化数据 5, 7, 15, 28—30, 32

structured query language 结构化查询语言 (SQL) 29, 33

supervised learning algorithm 有监督机器学习算法 20, 24

Sweeney, Latanya 拉坦娅·斯威妮 73

T

Target retail store 塔吉特零售店 91

targeted advertising 定向广告 8, 78—80, 85

Tesla 特斯拉 10, 11

Thucydides 修昔底德 1, 19

TOR(The Onion Router) 洋葱路由 102—103

Twitter 推特(网站)8, 17, 31, 52, 60, 65, 78, 100

U

Uconnect U 链接(汽车电脑)107

Ulbricht, Ross William 罗斯·威廉·乌布利希 103

大
数
据

120

Uniform Resource Locator 网络地址 (URL) 74

United States Geological Survey 美国地质调查局 (USGS) 13

unstructured data 非结构化数据 5, 6, 30, 31, 32, 60, 71, 78

unsupervised learning algorithm 无监督机器学习算法 20, 21

Upper Paleolithic era 旧石器时代晚期 2

US National Security Agency 美国国家安全局 (NSA) 97—99

V

Valasek, Chris 克里斯·瓦拉塞克 107

Valen, Snorre 斯诺里·瓦伦 101

variable selection 变量选择 63

variety 种类(多)10, 16—20, 59, 111

velocity 速度(快)16, 18, 20, 59

veracity 准确(性)18, 20, 59, 73

Vertesi, Janet 珍妮特·维尔特斯 102

vertical scalability 垂直扩展性 29

volume 数量(大)5, 7, 12, 16—17, 19, 20, 29, 59

Volvo 沃尔沃 10

W

warping compression (数据)扭曲压缩 42

Watson(IBM) "沃森" 医生 (IBM) 68, 70—72

Watson, James 詹姆斯·沃森 10

West Africa Ebola outbreak 西部非洲埃博拉暴发 65—67

wget program wget 程序 98

WikiLeaks 维基解密 99—102

Wired《连线》107

World Health Organization 世界卫生组织 (WHO) 65—67

World Wide Web (WWW) 万维网 17

Y

Yahoo! 雅虎(网站)7, 30, 78, 93, 104

Yoo, Ji Su 刘吉素 73

YouTube 优兔(网站)8

Z

zettabyte 十万亿亿字 (Zb) 37

Zika virus 寨卡病毒 67

索
引

Dawn E. Holmes

BIG DATA

A Very Short Introduction

Contents

Preface xv

Acknowledgements xvii

List of illustrations xix

1 The data explosion 1

2 Why is big data special? 14

3 Storing big data 26

4 Big data analytics 44

5 Big data and medicine 59

6 Big data, big business 75

7 Big data security and the Snowden case 90

8 Big data and society 105

Byte size chart 113

ASCII table for lower case letters 115

Further reading 117

Preface

Books on big data tend to fall into one of two categories: either they offer no explanation as to how things actually work or they are highly mathematical textbooks suitable only for graduate students. The aim of this book is to offer an alternative by providing an introduction to how big data works and is changing the world about us; the effect it has on our everyday lives; and the effect it has in the business world.

Data used to mean documents and papers, with maybe a few photos, but it now means much more than that. Social networking sites generate large amounts of data in the form of images, videos, and movies on a minute by minute basis. Online shopping creates data as we enter our address and credit card details. We are now at a point where the collection and storage of data is growing at a rate unimaginable only a few decades ago but, as we will see in this book, new data analysis techniques are transforming this data into useful information. While writing this book, I found that big data cannot be meaningfully discussed without frequent reference to its collection, storage, analysis, and use by the big commercial players. Since research departments in companies such as Google and Amazon have been responsible for many of the major developments in big data, frequent reference will be made to them.

The first chapter introduces the reader to the diversity of data in general before explaining how the digital age has led to changes in the way we define data. Big data is introduced informally through the idea of the data explosion, which involves computer science, statistics, and the interface between them. In Chapters 2 to 4, I have used diagrams quite extensively to help explain some of the new methods required by big data. The second chapter explores what makes big data special and, in doing so, leads us to a more specific definition. In Chapter 3, we discuss the problems related to storing and managing big data. Most people are familiar with the need to back up the data on their personal computer. But how do we do this with the colossal amounts of data that are now being generated? To answer this question, we will look at database storage and the idea of distributing tasks across clusters of computers. Chapter 4 argues that big data is only useful if we can extract useful information from it. A flavour of how data is turned into information is given using simplified explanations of several well-established techniques.

We then move on to a more detailed discussion of big data applications, starting in Chapter 5 with the role of big data in medicine. Chapter 6 analyses business practices with case studies on Amazon and Netflix, each highlighting different features of marketing using big data. Chapter 7 looks at some of the security issues surrounding big data and the importance of encryption. Data theft has become a big problem and we look at some of the cases that have been in the news including Snowden and WikiLeaks. The chapter concludes by showing how cybercrime is an issue that big data needs to address. In the final chapter, Chapter 8, we consider how big data is changing the society we live in, through the development of sophisticated robots and their role in the workplace. A consideration of the smart homes and smart cities of the future concludes the book.

In a very short introduction it is not possible to mention everything, so I hope the reader will pursue their interests through the Further reading section's recommendations.

Acknowledgements

When I mentioned to Peter that I wanted to acknowledge his contribution to this book, he suggested the following: 'I would like to thank Peter Harper, without whose dedicated use of the spell-checker this would have been a different book.' I would also thank him for his expertise in coffee-making and his sense of humour! By itself, this support is invaluable but Peter did much, much more and it is true to say that without his unwavering encouragement and constructive contributions, this book would not have been written.

Dawn E. Holmes

April 2017

List of illustrations

1 A cluster diagram **23**

2 Fraud dataset with known classifications **24**

3 Decision tree for transactions **25**

4 Simplified view of part of a Hadoop DFS cluster **31**

5 Key–value database **35**

6 Graph database **35**

7 A coded character string **38**

8 A binary tree **39**

9 The binary tree with a new node **40**

10 Completed binary tree **40**

11 Map function **46**

12 Shuffle and reduce functions **47**

13 10-bit array **49**

14 Summary of hash function results **50**

15 Bloom filter for malicious email addresses **50**

16 Directed graph representing a small part of the Web **53**

17 Directed graph representing a small part of the Web with added link **54**

18 Votes cast for each webpage **55**

19 Books bought by Smith, Jones, and Brown **81**

20 Jaccard index and distance **82**

21 Star ratings for purchases **82**

Chapter 1
The data explosion

What is data?

In 431 BCE, Sparta declared war on Athens. Thucydides, in his account of the war, describes how besieged Plataean forces loyal to Athens planned to escape by scaling the wall surrounding Plataea built by Spartan-led Peloponnesian forces. To do this they needed to know how high the wall was so that they could make ladders of suitable length. Much of the Peloponnesian wall had been covered with rough pebbledash, but a section was found where the bricks were still clearly visible and a large number of soldiers were each given the task of counting the layers of these exposed bricks. Working at a distance safe from enemy attack inevitably introduced mistakes, but as Thucydides explains, given that many counts were taken, the result that appeared most often would be correct. This most frequently occurring count, which we would now refer to as *the mode*, was then used to calculate the height of the wall, the Plataeans knowing the size of the local bricks used, and ladders of the length required to scale the wall were constructed. This enabled a force of several hundred men to escape, and the episode may well be considered the most impressive example of historic data collection and analysis. But the collection, storage, and analysis of data pre-dates even Thucydides by many centuries, as we will see.

Notches have been found on sticks, stones, and bones dating back to as early as the Upper Paleolithic era. These notches are thought to represent data stored as tally marks, though this is still open to academic debate. Perhaps the most famous example is the Ishango Bone, found in the Democratic Republic of Congo in 1950, and which is estimated to be around 20,000 years old. This notched bone has been variously interpreted as a calculator or a calendar, although others prefer to explain the notches as being there just to provide a better grip. The Lebombo Bone, discovered in the 1970s in Swaziland, is even older, dating from around 35,000 BCE. With twenty-nine lines scored across it, this fragment of a baboon's fibula bears a striking resemblance to the calendar sticks still used by bushmen in distant Namibia, suggesting that this may indeed be a method that was used to keep track of data important to their civilization.

While the interpretation of these notched bones is still open to speculation, we know that one of the first well-documented uses of data is the census undertaken by the Babylonians in 3800 BCE. This census systematically documented population numbers and commodities, such as milk and honey, in order to provide the information necessary to calculate taxes. The early Egyptians also used data, in the form of hieroglyphs written on wood or papyrus, to record the delivery of goods and to keep track of taxes. But early examples of data usage are by no means confined to Europe and Africa. The Incas and their South American predecessors, keen to record statistics for tax and commercial purposes, used a sophisticated and complex system of coloured knotted strings, called *quipu*, as a decimal-based accounting system. These knotted strings, made from brightly dyed cotton or camelid wool, date back to the third millennium BCE, and although fewer than a thousand are known to have survived the Spanish invasion and subsequent attempt to eradicate them, they are among the first known examples of a massive data storage system. Computer algorithms are now being developed to try to decode the full meaning of the *quipu* and enhance our understanding of how they were used.

Although we can think of and describe these early systems as using data, the word 'data' is actually a plural word of Latin origin, with 'datum' being the singular. 'Datum' is rarely used today and 'data' is used for both singular and plural. The *Oxford English Dictionary* attributes the first known use of the term to the 17th-century English cleric Henry Hammond in a controversial religious tract published in 1648. In it Hammond used the phrase 'heap of data', in a theological sense, to refer to incontrovertible religious truths. But although this publication stands out as representing the first use of the term 'data' in English, it does not capture its use in the modern sense of denoting facts and figures about a population of interest. 'Data', as we now understand the term, owes its origins to the scientific revolution in the 18th century led by intellectual giants such as Priestley, Newton, and Lavoisier; and, by 1809, following the work of earlier mathematicians, Gauss and Laplace were laying the highly mathematical foundations for modern statistical methodology.

On a more practical level, an extensive amount of data was collected on the 1854 cholera outbreak in Broad Street, London, allowing physician John Snow to chart the outbreak. By doing so, he was able to lend support to his hypothesis that contaminated water spread the disease and to show that it was not airborne as had been previously believed. Gathering data from local inhabitants he established that those affected were all using the same public water pump; he then persuaded the local parish authorities to shut it down, a task they accomplished by removing the pump handle. Snow subsequently produced a map, now famous, showing that the illness had occurred in clusters around the Broad Street pump. He continued to work in this field, collecting and analysing data, and is renowned as a pioneering epidemiologist.

Following John Snow's work, epidemiologists and social scientists have increasingly found demographic data invaluable for research purposes, and the census now taken in many countries proves a useful source of such information. For example, data on the birth and death rate, the frequency of various diseases, and statistics on income and crime is all now collected, which was not the case prior to the 19th century. The census, which takes place every ten years in most countries, has been collecting increasing amounts of data, which eventually has resulted in more than could realistically be recorded by hand or the simple tallying machines previously used. The challenge of processing these ever-increasing amounts of census data was in part met by Herman Hollerith while working for the US Census Bureau.

By the 1870 US census, a simple tallying machine was in operation but this had limited success in reducing the work of the Census Bureau. A breakthrough came in time for the 1890 census, when Herman Hollerith's punched cards tabulator for storing and processing data was used. The time taken to process the US census data was usually about eight years, but using this new invention the time was reduced to one year. Hollerith's machine revolutionized the analysis of census data in countries worldwide, including Germany, Russia, Norway, and Cuba.

Hollerith subsequently sold his machine to the company that evolved into IBM, which then developed and produced a widely used series of punch card machines. In 1969, the American National Standards Institute (ANSI) defined the Hollerith Punched Card Code (or Hollerith Card Code), honouring Hollerith for his early punch card innovations.

Data in the digital age

Before the widespread use of computers, data from the census, scientific experiments, or carefully designed sample surveys and questionnaires was recorded on paper—a process that was time-consuming and expensive. Data collection could only take place once researchers had decided which questions they wanted their experiments or surveys to answer, and the resulting highly structured data, transcribed onto paper in ordered rows and columns, was then amenable to traditional methods of statistical analysis. By the first half of the 20th century some data was being stored on computers, helping to alleviate some of this labour-intensive work, but it was through the launch of the World Wide Web (or Web) in 1989, and its rapid development, that it became increasingly feasible to generate, collect, store, and analyse data electronically. The problems inevitably generated by the very large volume of data made accessible by the Web then needed to be addressed, and we first look at how we may make distinctions between different types of data.

The data we derive from the Web can be classified as structured, unstructured, or semi-structured.

Structured data, of the kind written by hand and kept in notebooks or in filing cabinets, is now stored electronically on spreadsheets or databases, and consists of spreadsheet-style tables with rows and columns, each row being a record and each column a well-defined field (e.g. name, address, and age). We are contributing to these structured data stores when, for example, we provide the information necessary to order goods online. Carefully structured and tabulated data is relatively easy to manage and is amenable to statistical analysis, indeed until recently statistical analysis methods could be applied only to structured data.

In contrast, unstructured data is not so easily categorized and includes photos, videos, tweets, and word-processing documents. Once the use of the World Wide Web became widespread, it transpired that many such potential sources of information remained inaccessible because they lacked the structure needed for existing analytic techniques to be applied. However, by identifying key features, data that appears at first sight to be unstructured may not be completely without structure. Emails, for example, contain structured *metadata* in the heading as well as the actual unstructured message in the text and so may be classified as semi-structured data. Metadata tags, which are essentially descriptive references, can be used to add some structure to unstructured data. Adding a word tag to an image on a website makes it identifiable and so easier to search for. Semi-structured data is also found on social networking sites, which use hashtags so that messages (which are unstructured data) on a particular topic can be identified. Dealing with unstructured data is challenging: since it cannot be stored in traditional databases or spreadsheets, special tools have had to be developed to extract useful information. In later chapters we will look at how unstructured data is stored.

The term 'data explosion', which heads this chapter, refers to the increasingly vast amounts of structured, unstructured, and semi-structured data being generated minute by minute; we will look next at some of the many different sources that produce all this data.

Introduction to big data

Just in researching material for this book I have been swamped by the sheer volume of data available on the Web—from websites, scientific journals, and e-textbooks. According to a recent worldwide study conducted by IBM, about 2.5 *exabytes* (Eb) of data are generated every day. One Eb is 10^{18} (1 followed by eighteen 0s) bytes (or a million *terabytes* (Tb); see the Big data byte size chart at the end of this book). A good laptop bought at the time of writing will typically have a hard drive with 1 or 2 Tb of storage space. Originally, the term 'big data' simply referred to the very large amounts of data being produced in the digital age. These huge amounts of data, both structured and unstructured, include all the Web data generated by emails, websites, and social networking sites.

Approximately 80 per cent of the world's data is unstructured in the form of text, photos, and images, and so it is not amenable to the traditional methods of structured data analysis. 'Big data' is now used to refer not just to the total amount of data generated and stored electronically, but also to specific datasets that are large in both size and complexity, with which new algorithmic techniques are required in order to extract useful information from them. These big datasets come from different sources so let's take a more detailed look at some of them and the data they generate.

Search engine data

In 2015, Google was by far the most popular search engine worldwide, with Microsoft's Bing and Yahoo Search coming second and third, respectively. In 2012, the most recent year for which data is publicly available, there were over 3.5 billion searches made per day on Google alone.

Entering a key term into a search engine generates a list of the most relevant websites, but at the same time a considerable amount of data is being collected. Web tracking generates big data. As an exercise, I searched on 'border collies' and clicked on the top website returned. Using some basic tracking software, I found that some sixty-seven third-party site connections were generated just by clicking on this one website. In order to track the interests of people who access the site, information is being shared in this way between commercial enterprises.

Every time we use a search engine, logs are created recording which of the recommended sites we visited. These logs contain useful information such as the query term itself, the IP address of the device used, the time when the query was submitted, how long we stayed on each site, and in which order we visited them—all without identifying us by name. In addition, *clickstream logs* record the path taken as we visit various websites as well as our navigation within each website. When we surf the Web, every click we make is recorded somewhere for future use. Software is available for businesses allowing them to collect the clickstream data generated by their own website—a valuable marketing tool. For example, by providing data on the use of the system, logs can help detect malicious activity such as identity theft. Logs are also used to gauge the effectiveness of online advertising, essentially by counting the number of times an advertisement is clicked on by a website visitor.

By enabling customer identification, cookies are used to personalize your surfing experience. When you make your first visit to a chosen website, a *cookie*, which is a small text file, usually consisting of a website identifier and a user identifier, will be sent to your computer, unless you have blocked the use of cookies. Each time you visit this website, the cookie sends a message back to the website and in this way keeps track of your visits. As we will see in Chapter 6, cookies are often used to record clickstream data, to keep track of your preferences, or to add your name to targeted advertising.

Social networking sites also generate a vast amount of data, with Facebook and Twitter at the top of the list. By the middle of 2016, Facebook had, on average, 1.71 billion active users per month, all generating data, resulting in about 1.5 *petabytes* (Pb; or 1,000 Tb) of Web log data every day. YouTube, the popular video-sharing website, has had a huge impact since it started in 2005, and a recent YouTube press release claims that there are over a billion users worldwide. The valuable data produced by search engines and social networking sites can be used in many other areas, for example when dealing with health issues.

Healthcare data

If we look at healthcare we find an area which involves a large and growing percentage of the world population and which is increasingly computerized. Electronic health records are gradually becoming the norm in hospitals and doctors' surgeries, with the primary aim being to make it easier to share patient data with other hospitals and physicians, and so to facilitate the provision of better healthcare. The collection of personal data through wearable or implantable sensors is on the increase, particularly for health monitoring, with many of us using personal fitness trackers of varying complexity which output ever more categories of data. It is now possible to monitor a patient's health remotely in real-time through the collection of data on blood pressure, pulse, and temperature, thus potentially reducing healthcare costs and improving quality of life. These remote monitoring devices are becoming increasingly sophisticated and now go beyond basic measurements to include sleep tracking and arterial oxygen saturation rate.

Some companies offer incentives in order to persuade employees to use a wearable fitness device and to meet certain targets such as weight loss or a certain number of steps taken per day. In return for being given the device, the employee agrees to share the data with the employer. This may seem reasonable but there will inevitably be privacy issues to be considered, together with the unwelcome pressure some people may feel under to opt into such a scheme.

Other forms of employee monitoring are becoming more frequent, such as tracking all employee activities on the company-provided computers and smartphones. Using customized software, this tracking can include everything from monitoring which websites are visited to logging individual keystrokes and checking whether the computer is being used for private purposes such as visiting social network sites. In the age of massive data leaks, security is of growing concern and so corporate data must be protected. Monitoring emails and tracking websites visited are just two ways of reducing the theft of sensitive material.

As we have seen, personal health data may be derived from sensors, such as a fitness tracker or health monitoring device. However, much of the data being collected from sensors is for highly specialized medical purposes. Some of the largest data stores in existence are being generated as researchers study the genes and sequencing genomes of a variety of species. The structure of the deoxyribonucleic acid molecule (DNA), famous for holding the genetic instructions for the functioning of living organisms, was first described as a double-helix by James Watson and Francis Crick in 1953. One of the most highly publicized research projects in recent years has been the international human genome project, which determines the sequence, or exact order, of the three billion base-pairs that comprise human DNA. Ultimately, this data is helping research teams in the study of genetic diseases.

Real-time data

Some data is collected, processed, and used in real-time. The increase in computer processing power has allowed an increase in the ability to process as well as generate such data rapidly. These are systems where response time is crucial and so data must be processed in a timely manner. For example, the Global Positioning System (GPS) uses a system of satellites to scan the Earth and send back huge amounts of real-time data. A GPS receiving device, maybe in your car or smartphone ('smart' indicates that an item, in this case a phone, has Internet access and the ability to provide a number of services or applications (apps) that can then be linked together), processes these satellite signals and calculates your position, time, and speed.

This technology is now being used in the development of driverless or autonomous vehicles. These are already in use in confined, specialized areas such as factories and farms, and are being developed by a number of major manufacturers, including Volvo, Tesla, and Nissan. The sensors and computer programs involved have to process data in real-time to reliably navigate to your destination and control movement of the vehicle in relation to other road users. This involves prior creation of 3D maps of the routes to be used since the sensors cannot cope with non-mapped routes. Radar sensors are used to monitor other traffic, sending back data to an external central executive computer which controls the car. Sensors have to be programmed to detect shapes and distinguish between, for example, a child running into the road and a newspaper blowing across it; or to detect, say, an emergency traffic layout following an accident. However, these cars do not yet have the ability to react appropriately to all the problems posed by an ever-changing environment.

The first fatal crash involving an autonomous vehicle occurred in 2016, when neither the driver nor the autopilot reacted to a vehicle cutting across the car's path, meaning that the brakes were not applied. Tesla, the makers of the autonomous vehicle, in a June 2016 press release referred to the 'extremely rare circumstances of the impact'. The autopilot system warns drivers to keep their hands on the wheel at all times and even checks that they are doing so. Tesla state that this is the first fatality linked to their autopilot in 130 million miles of driving, compared with one fatality per 94 million miles of regular, non-automated driving in the US.

It has been estimated that each autonomous car will generate on average 30 Tb of data daily, much of which will have to be processed almost instantly. A new area of research, called *streaming analytics*, which bypasses traditional statistical and data processing methods, hopes to provide the means for dealing with this particular big data problem.

Astronomical data

In April 2014 an International Data Corporation report estimated that, by 2020, the digital universe will be 44 trillion *gigabytes*

The data explosion

(Gb; or 1,000 *megabytes* (Mb)), which is about ten times its size in 2013. An increasing volume of data is being produced by telescopes. For example, the Very Large Telescope in Chile is an optical telescope, which actually consists of four telescopes, each producing huge amounts of data—15 Tb per night, every night in total. It will spearhead the Large Synoptic Survey, a ten-year project repeatedly producing maps of the night sky, creating an estimated grand total of 60 Pb (2^{50} bytes).

Even bigger in terms of data generation is the Square Kilometer Array Pathfinder (ASKAP) radio telescope being built in Australia and South Africa, which is projected to begin operation in 2018. It will produce 160 Tb of raw data per second initially, and ever more as further phases are completed. Not all this data will be stored but even so, supercomputers around the world will be needed to analyse the remaining data.

What use is all this data?

It is now almost impossible to take part in everyday activities and avoid having some personal data collected electronically. Supermarket check-outs collect data on what we buy; airlines collect information about our travel arrangements when we purchase a ticket; and banks collect our financial data.

Big data is used extensively in commerce and medicine and has applications in law, sociology, marketing, public health, and all areas of natural science. Data in all its forms has the potential to provide a wealth of useful information if we can develop ways to extract it. New techniques melding traditional statistics and computer science make it increasingly feasible to analyse large sets of data. These techniques and algorithms developed by statisticians and computer scientists search for patterns in data. Determining which patterns are important is key to the success of big data analytics. The changes brought about by the digital age have substantially changed the way data is collected, stored, and analysed. The big data revolution has given us smart cars and home-monitoring.

The ability to gather data electronically resulted in the emergence of the exciting field of data science, bringing together the disciplines of statistics and computer science in order to analyse these large quantities of data to discover new knowledge in interdisciplinary areas of application. The ultimate aim of working with big data is to extract useful information. Decision-making in business, for example, is increasingly based on the information gleaned from big data, and expectations are high. But there are significant problems, not least with the shortage of trained data scientists capable of effectively developing and managing the systems necessary to extract the desired information.

By using new methods derived from statistics, computer science, and artificial intelligence, algorithms are now being designed that result in new insights and advances in science. For example, although it is not possible to predict exactly when and where an earthquake will occur, an increasing number of organizations are using data collected by satellite and ground sensors to monitor seismic activity. The aim is to determine approximately where big earthquakes are *likely* to occur in the long-term. For example, the US Geological Survey (USGS), a major player in seismic research, estimated in 2016 that 'there is a 76% probability that a magnitude 7 earthquake will occur within the next 30 years in northern California'. Probabilities such as these help focus resources on measures such as ensuring that buildings are better able to withstand earthquakes and having disaster management programmes in place. Several companies in these and other areas are working with big data to provide improved forecasting methods, which were not available before the advent of big data. We need to take a look at what is special about big data.

Chapter 2
Why is big data special?

Big data didn't just happen—it was closely linked to the development of computer technology. The rapid rate of growth in computing power and storage led to progressively more data being collected, and, regardless of who first coined the term, 'big data' was initially all about size. Yet it is not possible to define big data exclusively in terms of how many Pb, or even Eb, are being generated and stored. However, a useful means for talking about the 'big data' resulting from the data explosion is provided by the term 'small data'—although it is not widely used by statisticians. Big datasets are certainly large and complex, but in order for us to reach a definition, we need first to understand 'small data' and its role in statistical analysis.

Big data versus small data

In 1919, Ronald Fisher, now widely recognized as the founder of modern statistics as an academically rigorous discipline, arrived at Rothamsted Agricultural Experimental Station in the UK to work on analysing crop data. Data has been collected from the Classical Field Experiments conducted at Rothamsted since the 1840s, including both their work on winter wheat and spring barley and meteorological data from the field station. Fisher started the Broadbalk project which examined the effects of different fertilizers on wheat, a project still running today.

Recognizing the mess the data was in, Fisher famously referred to his initial work there as 'raking over the muck heap'. However, by meticulously studying the experimental results that had been carefully recorded in leather-bound note books he was able to make sense of the data. Working under the constraints of his time, before today's computing technology, Fisher was assisted only by a mechanical calculator as he, nonetheless successfully, performed calculations on seventy years of accumulated data. This calculator, known as the Millionaire, which relied for power on a tedious hand-cranking procedure, was innovative in its day, since it was the first commercially available calculator that could be used to perform multiplication. Fisher's work was computationally intensive and the Millionaire played a crucial role in enabling him to perform the many required calculations that any modern computer would complete within seconds.

Although Fisher collated and analysed a lot of data it would not be considered a large amount today, and it would certainly not be considered 'big data'. The crux of Fisher's work was the use of precisely defined and carefully controlled experiments, designed to produce highly structured, unbiased sample data. This was essential since the statistical methods then available could only be applied to structured data. Indeed, these invaluable techniques still provide the cornerstone for the analysis of small, structured sets of data. However, those techniques are not applicable to the very large amounts of data we can now access with so many different digital sources available to us.

Big data defined

In the digital age we are no longer entirely dependent on samples, since we can often collect all the data we need on entire populations. But the size of these increasingly large sets of data cannot alone provide a definition for the term 'big data'—we must include *complexity* in any definition. Instead of carefully constructed samples of 'small data' we are now dealing with huge amounts of data that has not been collected with any specific questions in mind and is often unstructured. In order to characterize the key features that make data big and move towards a definition of the term, Doug Laney, writing in 2001, proposed using the three 'v's: volume, variety, and velocity. By looking at each of these in turn we can get a better idea of what the term 'big data' means.

Volume

'Volume' refers to the amount of electronic data that is now collected and stored, which is growing at an ever-increasing rate. Big data is big, but how big? It would be easy just to set a specific size as denoting 'big' in this context, but what was considered 'big' ten years ago is no longer big by today's standards. Data acquisition is growing at such a rate that any chosen limit would inevitably soon become outdated. In 2012, IBM and the University of Oxford reported the findings of their Big Data Work Survey. In this international survey of 1,144 professionals working in ninety-five different countries, over half judged datasets of between 1 Tb and 1 Pb to be big, while about a third of respondents fell in the 'don't know' category. The survey asked respondents to choose either one or two defining characteristics of big data from a choice of eight; only 10 per cent voted for 'large volumes of data' with the top choice being 'a greater scope of information', which attracted 18 per cent. Another reason why there can be no definitive limit based solely on size is because other factors, like storage and the type of data being collected, change over time and affect our perception of volume. Of course, some datasets are very big indeed, including, for example, those obtained by the Large Hadron Collider at CERN, the world's premier particle accelerator, which has been operating since 2008. Even after extracting only 1 per cent of the total data generated, scientists still have 25 Pb to process annually. Generally, we can say the volume criterion is met if the dataset is such that we cannot collect, store, and analyse it using traditional computing and statistical methods. Sensor data, such as that generated by the Large Hadron Collider, is just one variety of big data, so let's consider some of the others.

Variety

Though you may often see the terms 'Internet' and 'World Wide Web' used interchangeably, they are actually very different. The Internet is a network of networks, consisting of computers, computer networks, local area networks (LANs), satellites, and cellphones and other electronic devices, all linked together and able to send bundles of data to one another, which they do using an IP (Internet protocol) address. The World Wide Web (www, or Web), described by its inventor, T. J. Berners-Lee, as 'a global information system', exploited Internet access so that all those with a computer and a connection could communicate with other users through such media as email, instant messaging, social networking, and texting. Subscribers to an ISP (Internet services provider) can connect to the Internet and so access the Web and many other services.

Once we are connected to the Web, we have access to a chaotic collection of data, from sources both reliable and suspect, prone to repetition and error. This is a long way from the clean and precise data demanded by traditional statistics. Although the data collected from the Web can be structured, unstructured, or semi-structured resulting in significant variety (e.g. unstructured word-processed documents or posts found on social networking sites; and semi-structured spreadsheets), most of the big data derived from the Web is unstructured. Twitter users, for example, publish approximately 500 million 140-character messages, or *tweets*, per day worldwide. These short messages are valuable commercially and are often analysed according to whether the sentiment expressed is positive, negative, or neutral. This new area of sentiment analysis requires specially developed techniques and is something we can do effectively only by using big data analytics. Although a great variety of data is collected by hospitals, the military, and many commercial enterprises for a number of purposes, ultimately it can all be classified as structured, unstructured, or semi-structured.

Velocity

Data is now streaming continuously from sources such as the Web, smartphones, and sensors. Velocity is necessarily connected with volume: the faster data is generated, the more there is. For example, the messages on social media that now 'go viral' are transmitted in such a way as to have a snowball effect: I post something on social media, my friends look at it, and each shares it with their friends, and so on. Very quickly these messages make their way around the world.

Velocity also refers to the speed at which data is electronically processed. For example, sensor data, such as that being generated by an autonomous car, is necessarily generated in real-time. If the car is to work reliably, the data, sent wirelessly to a central location, must be analysed very quickly so that the necessary instructions can be sent back to the car in a timely fashion.

Variability may be considered as an additional dimension of the velocity concept, referring to the changing rates in flow of data, such as the considerable increase in data flow during peak times. This is significant because computer systems are more prone to failure at these times.

Veracity

As well as the original three 'v's suggested by Laney, we may add 'veracity' as a fourth. Veracity refers to the quality of the data being collected. Data that is accurate and reliable has been the hallmark of statistical analysis in the past century. Fisher, and others, strived to devise methods encapsulating these two concepts, but the data generated in the digital age is often unstructured, and often collected without experimental design or, indeed, any concept of what questions might be of interest. And yet we seek to gain information from this mish-mash. Take, for example, the data generated by social networks. This data is by its very nature imprecise, uncertain, and often the information posted is simply not true. So how can we trust the data to yield meaningful results? Volume can help in overcoming these problems—as we saw in Chapter 1, when Thucydides described the Plataean forces engaging the greatest possible number of soldiers counting bricks in order to be more likely to get (close to) the correct height of the wall they wished to scale. However, we need to be more cautious, as we know from statistical theory, greater volume can lead to the opposite result, in that, given sufficient data, we can find any number of spurious correlations.

Visualization and other 'v's

'V' has become the letter of choice, with competing definitions adding or substituting such terms as 'vulnerability' and 'viability' to Laney's original three—the most important perhaps of these additions being 'value' and 'visualization'. Value generally refers to the quality of the results derived from big data analysis. It has also been used to describe the selling by commercial enterprises of data to firms who then process it using their own analytics, and so it is a term often referred to in the data business world.

Visualization is not a characterizing feature of big data, but it is important in the presentation and communication of analytic results. The familiar static pie charts and bar graphs that help us to understand small datasets have been further developed to aid in the visual interpretation of big data, but these are limited in their applicability. Infographics, for example, provide a more complex presentation but are static. Since big data is constantly being added to, the best visualizations are interactive for the user and updated regularly by the originator. For example, when we

use GPS for planning a car journey, we are accessing a highly
interactive graphic, based on satellite data, to track our position.

Taken together, the four main characteristics of big data—volume,
variety, velocity, and veracity—present a considerable challenge
in data management. The advantages we expect to gain from
meeting this challenge and the questions we hope to answer with
big data can be understood through data mining.

Big data mining

'Data is the new oil', a phrase that is common currency among
leaders in industry, commerce, and politics, is usually attributed to
Clive Humby in 2006, the originator of Tesco's customer loyalty
card. It's a catchy phrase and suggests that data, like oil, is
extremely valuable but must first be processed before that value
can be realized. The phrase is primarily used as a marketing
ploy by data analytics providers hoping to sell their products by
convincing companies that big data is the future. It may well be,
but the metaphor only holds so far. Once you strike oil you have a
marketable commodity. Not so with big data; unless you have the
right data you can produce nothing of value. Ownership is an
issue; privacy is an issue; and, unlike oil, data appears not to be a
finite resource. However, continuing loosely with the industrial
metaphor, mining big data is the task of extracting useful and
valuable information from massive datasets.

Using data mining and machine learning methods and algorithms,
it is possible not only to detect unusual patterns or anomalies in
data, but also to predict them. In order to acquire this kind of
knowledge from big datasets, either supervised or unsupervised
machine learning techniques may be used. Supervised machine
learning can be thought of as roughly comparable to learning from
example in humans. Using training data, where correct examples
are labelled, a computer program develops a rule or algorithm for
classifying new examples. This algorithm is checked using the test

data. In contrast, unsupervised learning algorithms use unlabelled input data and no target is given; they are designed to explore data and discover hidden patterns.

As an example let's look at credit card fraud detection, and see how each method is used.

Credit card fraud detection

A lot of effort goes into detecting and preventing credit card fraud. If you have been unfortunate enough to receive a phone call from your credit card fraud detection office, you may be wondering how the decision was reached that the recently made purchase on your card had a good chance of being fraudulent. Given the huge number of credit card transactions it is no longer feasible to have humans checking transactions using traditional data analysis techniques, and so big data analytics are increasingly becoming necessary. Understandably, financial institutions are unwilling to share details of their fraud detection methods since doing so would give cyber criminals the information they need to develop ways round it. However, the broad brush strokes present an interesting picture.

There are several possible scenarios but we can look at personal banking and consider the case in which a credit card has been stolen and used in conjunction with other stolen information, such as the card PIN (personal identification number). In this case, the card might show a sudden increase in expenditure—a fraud that is easily detected by the card issuing agency. More often, a fraudster will first use a stolen card for a 'test transaction' in which something inexpensive is purchased. If this does not raise any alarms, then a bigger amount is taken. Such transactions may or may not be fraudulent—maybe a cardholder bought something outside of their usual purchasing pattern, or maybe they actually just spent a lot that month. So how do we detect which transactions are fraudulent? Let's look first at an unsupervised technique, called *clustering*, and how it might be used in this situation.

Clustering

Based on artificial intelligence algorithms, clustering methods can be used to detect anomalies in customer purchasing behaviour. We are looking for patterns in transaction data and want to detect anything unusual or suspicious which may or may not be fraudulent.

A credit card company gathers lots of data and uses it to form profiles showing the purchasing behaviour of their customers. Clusters of profiles with similar properties are then identified electronically using an *iterative* (i.e. repeating a process to generate a result) computer program. For example, a cluster may be defined on accounts with a typical spending range or location, a customer's upper spending limit, or on the kind of items purchased, each resulting in a separate cluster.

When data is collected by a credit card provider it does not carry any label indicating whether the transactions are genuine or fraudulent. Our task is to use this data as input and, using a suitable algorithm, accurately categorize transactions. To do this, we will need to find similar groups, or clusters, within the input data. So, for example, we might group data according to the amount spent, the location where the transaction took place, the kind of purchase made, or the age of the card holder. When a new transaction is made, the cluster identification is computed for that transaction and if it is different from the existing cluster identification for that customer, it is treated as suspicious. Even if it falls within the usual cluster, if it is sufficiently far from the centre of the cluster it may still arouse suspicion.

For example, say an 83-year-old grandmother living in Pasadena purchases a flashy sports car; if this does not cluster with her

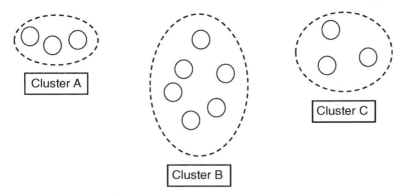

Cluster A

Cluster B

Cluster C

1. A cluster diagram.

usual purchasing behaviour of, say, groceries and visits to the hairdresser, it would be considered anomalous. Anything out of the ordinary, like this purchase, is considered worthy of further investigation, usually starting by contacting the card owner. In Figure 1 we see a very simple example of a cluster diagram representing this situation.

Cluster B shows the grandmother's usual monthly expenditure clustered with other people who have a similar monthly expenditure. Now, in some circumstances, for example when taking her annual vacation, the grandmother's expenditure for the month increases, perhaps grouping her with those in Cluster C, which is not too far distant from Cluster B and so not drastically dissimilar. Even so, since it is in a different cluster, it would be checked as suspicious account activity, but the purchase of the flashy sports car on her account puts her expenditure into Cluster A, which is very distant from her usual cluster and so is highly unlikely to reflect a legitimate purchase.

In contrast to this situation, if we already have a set of examples where we know fraud has occurred, instead of clustering algorithms we can use classification methods, which provide another data mining technique used for fraud detection.

Classification

Classification, a supervised learning technique, requires prior knowledge of the groups involved. We start with a dataset in which each observation is already correctly labelled or classified. This is divided into a *training set*, which enables us to build a classification model of the data, and a *test set*, which is used to check that the model is a good one. We can then use this model to classify new observations as they arise.

To illustrate classification, we will build a small decision tree for detecting credit card fraud.

To build our decision tree, let us suppose that credit card transaction data has been collected and transactions classified as genuine or fraudulent based on our historical knowledge are provided, as shown in Figure 2.

Using this data, we can build the decision tree shown in Figure 3, which will allow the computer to classify new transactions entering the system. We wish to arrive at one of the two possible transaction classifications, genuine or fraudulent, by asking a series of questions.

Was the card reported stolen or lost?	Was the item purchased unusual?	Was the customer phoned and asked if they made the purchase?	Classification
No	No		Genuine transaction
No	Yes	Yes	Genuine transaction
No	Yes	No	Fraudulent transaction
Yes			Fraudulent transaction

2. Fraud dataset with known classifications.

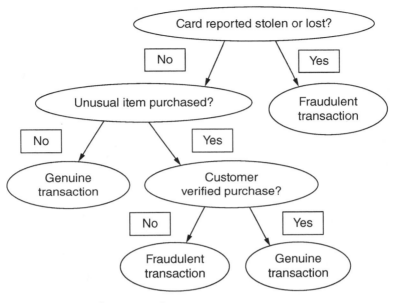

3. Decision tree for transactions.

By starting at the top of the tree in Figure 3, we have a series of test questions which will enable us to classify a new transaction.

For example, if Mr Smith's account shows that he has reported his credit card as lost or stolen, then any attempt to use it is deemed fraudulent. If the card has not been reported lost or stolen, then the system will check to see if an unusual item or an item costing an unusual amount for this customer has been purchased. If not, then the transaction is seen as nothing out of the ordinary and labelled as genuine. On the other hand, if the item is unusual then a phone call to Mr Smith will be triggered. If he confirms that he did make the purchase, then it is deemed genuine; if not, fraudulent.

Having arrived at an informal definition of big data, and considered the kinds of questions that can be answered by mining big data, let us now turn to the problems of data storage.

Chapter 3
Storing big data

The first hard drive, developed and sold by IBM in San Jose, California, had a storage capacity of about 5 Mb, held on fifty disks, each 24 inches in diameter. This was cutting edge technology in 1956. The device was physically massive, weighed over 1 ton, and was part of a mainframe computer. By the time of the Apollo 11 moon landing in 1969, NASA's Manned Spacecraft Center in Houston was using mainframe computers that each had up to 8 Mb of memory. Amazingly, the onboard computer for the Apollo 11 moon landing craft, piloted by Neil Armstrong, had a mere 64 *kilobytes* (Kb) of memory.

Computer technology progressed rapidly and by the start of the personal computer boom in the 1980s, the average hard drive on a PC was 5 Mb when one was included, which was not always the case. This would hold one or two photos or images today. Computer storage capacity increased very quickly and although personal computer storage has not kept up with big data storage, it has increased dramatically in recent years. Now, you can buy a PC with an 8 Tb hard drive or even bigger. Flash drives are now available with 1 Tb of storage, which is sufficient to store approximately 500 hours of movies or over 300,000 photos. This seems a lot until we contrast it with the estimated 2.5 Eb of new data being generated every day.

Once the change from valves to transistors took place in the 1960s the number of transistors that could be placed on a chip grew very rapidly, roughly in accordance with Moore's Law, which we discuss in the next section. And despite predictions that the limit of miniaturization was about to be reached it continues to be a reasonable and useful approximation. We can now cram billions of increasingly faster transistors onto a chip, which allows us to store ever greater quantities of data, while multi-core processors together with multi-threading software make it possible to process that data.

Moore's Law

In 1965, Gordon Moore, who became the co-founder of Intel, famously predicted that over the next ten years the number of transistors incorporated in a chip would approximately double every twenty-four months. In 1975, he changed his prediction and suggested the complexity would double every twelve months for five years and then fall back to doubling every twenty-four months. David House, a colleague at Intel, after taking into account the increasing speed of transistors, suggested that the *performance* of microchips would double every eighteen months, and it is currently the latter prediction that is most often used for Moore's Law. This prediction has proved remarkably accurate; computers have indeed become faster, cheaper, and more powerful since 1965, but Moore himself feels that this 'law' will soon cease to hold.

According to M. Mitchell Waldrop in an article published in the February 2016 edition of the scientific journal *Nature*, the end is indeed nigh for Moore's Law. A microprocessor is the integrated circuit responsible for performing the instructions provided by a computer program. This usually consists of billions of transistors, embedded in a tiny space on a silicon microchip. A gate in each transistor allows it to be either switched on or off and so it can be used to store 0 and 1. A very small input current flows through each transistor gate and produces an amplified output current when the gate is closed. Mitchell Waldrop was interested in the distance between gates, currently at 14-nanometer gaps in top microprocessors, and stated that the problems of heat generation caused by closer circuitry and how it is to be effectively dissipated were causing the exponential growth predicted by Moore's Law to falter, which drew our attention to the fundamental limits he saw rapidly approaching.

A nanometre is 10^{-9} metre, or one-millionth of a millimetre. To put this in context, a human hair is about 75,000 nanometres in diameter and the diameter of an atom is between 0.1 and 0.5 nanometres. Paolo Gargini, who works for Intel, claimed that the gap limit will be 2 or 3 nanometres and will be reached in the not too distant future—maybe as soon as the 2020s. Waldrop speculates that 'at that scale, electron behaviour will be governed by quantum uncertainties that will make transistors hopelessly unreliable'. As we will see in Chapter 7, it seems quite likely that quantum computers, a technology still in its infancy, will eventually provide the way forward.

Moore's Law is now also applicable to the rate of growth for data as the amount generated appears to approximately double every two years. Data increases as storage capacity increases and the capacity to process data increases. We are all beneficiaries: Netflix, smartphones, the Internet of Things (IoT; a convenient way of referring to the vast numbers of electronic sensors connected to the Internet), and the Cloud (a worldwide network of interconnected servers) computing, among others, have all become possible because of the exponential growth predicted by Moore's Law. All this generated data has to be stored, and we look at this next.

Storing structured data

Anyone who uses a personal computer, laptop, or smartphone accesses data stored in a database. Structured data, such as bank statements and electronic address books, are stored in a relational database. In order to manage all this structured data, a relational database management system (RDBMS) is used to create, maintain, access, and manipulate the data. The first step is to design the database schema (i.e. the structure of the database). In order to achieve this, we need to know the data fields and be able to arrange them in tables, and we then need to identify the relationships between the tables. Once this has been accomplished and the database constructed we can populate it with data and interrogate it using structured query language (SQL).

Clearly tables have to be designed carefully and it would require a lot of work to make significant changes. However, the relational model should not be underestimated. For many structured data applications, it is fast and reliable. An important aspect of relational database design involves a process called *normalization* which includes reducing data duplication to a minimum and hence reduces storage requirements. This allows speedier queries, but even so as the volume of data increases the performance of these traditional databases decreases.

The problem is one of scalability. Since relational databases are essentially designed to run on just one server, as more and more data is added they become slow and unreliable. The only way to achieve scalability is to add more computing power, which has its limits. This is known as *vertical scalability*. So although structured data is usually stored and managed in an RDBMS, when the data is big, say in terabytes or petabytes and beyond, the RDBMS no longer works efficiently, even for structured data.

An important feature of relational databases and a good reason for continuing to use them is that they conform to the following group of properties: atomicity, consistency, isolation, and durability, usually known as ACID. Atomicity ensures that incomplete transactions cannot update the database; consistency excludes invalid data; isolation ensures one transaction does not interfere with another transaction; and durability means that the database must update before the next transaction is carried out. All these are desirable properties but storing and accessing big data, which is mostly unstructured, requires a different approach.

Unstructured data storage

For unstructured data, the RDBMS is inappropriate for several reasons, not least that once the relational database schema has been constructed, it is difficult to change it. In addition, unstructured data cannot be organized conveniently into rows and columns. As we have seen, big data is often high-velocity and generated in real-time with real-time processing requirements, so although the RDBMS is excellent for many purposes and serves us well, given the current data explosion there has been intensive research into new storage and management techniques.

In order to store these massive datasets, data is distributed across servers. As the number of servers involved increases, the chance of failure at some point also increases, so it is important to have multiple, reliably identical copies of the same data, each stored on a different server. Indeed, with the massive amounts of data now being processed, systems failure is taken as inevitable and so ways of coping with this are built into the methods of storage. So how are the needs for speed and reliability to be met?

Hadoop Distributed File System

A distributed file system (DFS) provides effective and reliable storage for big data across many computers. Influenced by the ideas published in October 2003 by Google in a research paper launching the Google File System, Doug Cutting, who was then working at Yahoo, and his colleague Mike Cafarella, a graduate student at the University of Washington, went to work on developing the Hadoop DFS. Hadoop, one of the most popular

DFS, is part of a bigger, open-source software project called the Hadoop Ecosystem. Named after a yellow soft toy elephant owned by Cutting's son, Hadoop is written in the popular programming language, Java. If you use Facebook, Twitter, or eBay, for example, Hadoop will have been working in the background while you do so. It enables the storage of both semi-structured and unstructured data, and provides a platform for data analysis.

When we use Hadoop DFS, the data is distributed across many nodes, often tens of thousands of them, physically situated in data centres around the world. Figure 4 shows the basic structure of a single Hadoop DFS cluster, which consists of one master NameNode and many slave DataNodes.

The NameNode deals with all requests coming in from a client computer; it distributes storage space, and keeps track of storage availability and data location. It also manages all the basic file operations (e.g. opening and closing files) and controls data access by client computers. The DataNodes are responsible for actually storing the data and in order to do so, create, delete, and replicate blocks as necessary.

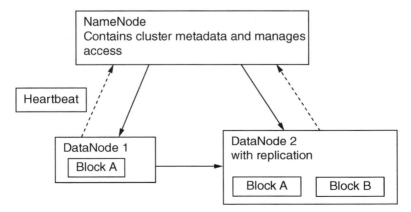

4. Simplified view of part of a Hadoop DFS cluster.

Data replication is an essential feature of the Hadoop DFS. For example, looking at Figure 4, we see that Block A is stored in both DataNode 1 and DataNode 2. It is important that several copies of each block are stored so that if a DataNode fails, other nodes are able to take over and continue with processing tasks without loss of data. In order to keep track of which DataNodes, if any, have failed, the NameNode receives a message from each, called a *Heartbeat*, every three seconds, and if no message is received it is assumed that the DataNode in question has ceased to function. So if DataNode 1 fails to send a Heartbeat, DataNode 2 will become the working node for Block A operations. The situation is different if the NameNode is lost, in which case the inbuilt backup system needs to be employed.

Data is written to a DataNode only once but will be read by an application many times. Each block is usually only 64 Mb, so there are a lot of them. One of the functions of the NameNode is to determine the best DataNode to use given the current usage, ensuring fast data access and processing. The client computer then accesses the data block from the chosen node. DataNodes are added as and when required by the increased storage requirements, a feature known as *horizontal scalability.*

One of the main advantages of Hadoop DFS over a relational database is that you can collect vast amounts of data, keep adding to it, and, at that time, not yet have any clear idea of what you want to use it for. Facebook, for example, uses Hadoop to store its continually growing amount of data. No data is lost, as it will store anything and everything in its original format. Adding DataNodes as required is cheap and does not require existing nodes to be changed. If previous nodes become redundant, it is easy to stop them working. As we have seen, structured data with identifiable rows and columns can be easily stored in a RDBMS while unstructured data can be stored cheaply and readily using a DFS.

NoSQL databases for big data

NoSQL is the generic name used to refer to non-relational databases and stands for *Not only SQL*. Why is there a need for a non-relational model that does not use SQL? The short answer is that the non-relational model allows us to continually add new data. The non-relational model has some features that are necessary in the management of big data, namely scalability, availability, and performance. With a relational database you cannot keep scaling vertically without loss of function, whereas with NoSQL you scale horizontally and this enables performance to be maintained. Before describing the NoSQL distributed database infrastructure and why it is suitable for big data, we need to consider the CAP Theorem.

CAP Theorem

In 2000, Eric Brewer, a professor of computer science at the University of California Berkeley, presented the CAP (consistency, availability, and partition tolerance) Theorem. Within the context of a distributed database system, consistency refers to the requirement that all copies of data should be the same across nodes. So in Figure 4, for example, Block A in DataNode 1 should be the same as Block A in DataNode 2. Availability requires that if a node fails, other nodes still function—if DataNode 1 fails, then DataNode 2 must still operate. Data, and hence DataNodes, are distributed across physically separate servers and communication between these machines will sometimes fail. When this occurs it is called a *network partition*. Partition tolerance requires that the system continues to operate even if this happens.

In essence, what the CAP Theorem states is that for any distributed computer system, where the data is shared, only two of these three criteria can be met. There are therefore three possibilities; the system must be: consistent and available, consistent and partition tolerant, or partition tolerant and available. Notice that since in a RDMS the network is not partitioned, only consistency and availability would be of concern and the RDMS model meets both of these criteria. In NoSQL, since we necessarily have partitioning, we have to choose between consistency and availability. By sacrificing availability, we are able to wait until consistency is achieved. If we choose instead to sacrifice consistency it follows that sometimes the data will differ from server to server.

The somewhat contrived acronym BASE (Basically Available, Soft, and Eventually consistent) is used as a convenient way of describing this situation. BASE appears to have been chosen in contrast to the ACID properties of relational databases. 'Soft' in this context refers to the flexibility in the consistency requirement. The aim is not to abandon any one of these criteria but to find a way of optimizing all three, essentially a compromise.

The architecture of NoSQL databases

The name NoSQL derives from the fact that SQL cannot be used to query these databases. So, for example, joins such as the one we saw in Figure 4 are not possible. There are four main types of non-relational or NoSQL database: key–value, column-based, document, and graph—all useful for storing large amounts of structured and semi-structured data. The simplest is the key–value database, which consists of an identifier (the *key*) and the data associated with that key (the *value*) as shown in Figure 5. Notice that 'value' can contain multiple items of data.

Of course, there would be many such key–value pairs and adding new ones or deleting old ones is simple enough, making the database highly scalable horizontally. The primary capability is that we can look up the value for a given key. For example, using the key 'Jane Smith' we are able to find her address. With huge amounts of data, this provides a fast, reliable, and readily scalable solution to storage

Key	Value
Jane Smith	Address: 33 Any Drive; Any City
Tom Brown	Gender: Male; Marital Status: Married; # of Children: 2; Favourite Movies: Cinderella; Dracula; Patton

5. **Key–value database.**

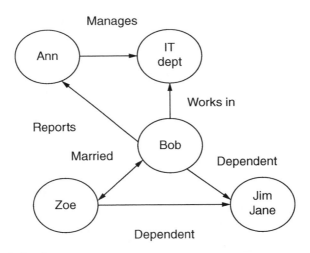

6. **Graph database.**

but it is limited by not having a query language. Column-based and document databases are extensions of the key–value model.

Graph databases follow a different model and are popular with social networking sites as well as being useful in business applications. These graphs are often very large, particularly when used by social networking sites. In this kind of database, the information is stored in nodes (i.e. vertices) and edges. For example, the graph in Figure 6 shows five nodes with the arrows between them representing relationships. Adding, updating, or deleting nodes changes the graph.

In this example, the nodes are names or departments, and the edges are the relationships between them. Data is retrieved from the graph by looking at the edges. So, for example, if I want to find 'names of employees in the IT department who have dependent children', I see that Bob fulfils both criteria. Notice that this is not a directed graph—we do not follow the arrows, we look for links.

Currently, an approach called NewSQL is finding a niche. By combining the performance of NoSQL databases and the ACID properties of the relational model, the aim of this latent technology is to solve the scalability problems associated with the relational model, making it more useable for big data.

Cloud storage

Like so many modern computing terms the Cloud sounds friendly, comforting, inviting, and familiar, but actually 'the Cloud' is, as mentioned earlier, just a way of referring to a network of interconnected servers housed in data centres across the world. These data centres provide a hub for storing big data.

Through the Internet we share the use of these remote servers, provided (on payment of a fee) by various companies, to store and manage our files, to run apps, and so on. As long as your computer or other device has the requisite software to access the Cloud, you can view your files from anywhere and give permission for others to do so. You can also use software that 'resides' in the Cloud rather than on your computer. So it's not just a matter of accessing the Internet but also of having the means to store and process information—hence the term 'Cloud computing'. Our individual Cloud storage needs are not that big, but scaled up the amount of information stored is massive.

Amazon is the biggest provider of Cloud services but the amount of data managed by them is a commercial secret. We can get some idea of their importance in Cloud computing by looking at an incident that occurred in February 2017 when Amazon Web Services' Cloud storage system, S3, suffered a major *outage* (i.e. service was lost). This lasted for approximately five hours and resulted in the loss of connection to many websites and services, including Netflix, Expedia, and the US Securities and Exchange Commission. Amazon later reported human error as the cause, stating that one of their employees had been responsible for inadvertently taking servers offline. Rebooting these large systems took longer than expected but was eventually completed successfully. Even so, the incident highlights the susceptibility of the Internet to failure, whether by a genuine mistake or by ill-intentioned hacking.

Lossless data compression

In 2017, the widely respected International Data Corporation (IDC) estimates that the digital universe totals a massive 16 *zettabytes* (Zb) which amounts to an unfathomable 16×10^{21} bytes. Ultimately, as the digital universe continues to grow, questions concerning what data we should actually save, how many copies should be kept, and for how long will have to be addressed. It rather challenges the *raison d'être* of big data to consider purging data stores on a regular basis or even archiving them, as this is in itself costly and potentially valuable data could be lost given that we do not necessarily know what data might be important to us in the future. However, with the huge amounts of data being stored, data compression has become necessary in order to maximize storage space.

There is considerable variability in the quality of the data collected electronically and so before it can be usefully analysed it must be pre-processed to check for and remedy problems with consistency, repetition, and reliability. Consistency is clearly important if we are to rely on the information extracted from the data. Removing unwanted repetitions is good housekeeping for any dataset, but with big datasets there is the additional concern that there may not be sufficient storage space available to keep all the data. Data is compressed to reduce redundancy in videos and images and so reduce storage requirements and, in the case of videos, to improve streaming rates.

There are two main types of compression—lossless and lossy. In *lossless compression* all the data is preserved and so this is particularly useful for text. For example, files with the extension '.ZIP', have been compressed without loss of information so that unzipping them returns us to the original file. If we compress the string of characters 'aaaaabbbbbbbbbb' as '5a10b' it is easy to see how to decompress and arrive at the original string. There are many algorithms for compression but it is useful first to consider how data is stored without compression.

ASCII (American Standard Code for Information Interchange) is the standard way of encoding data so that it can be stored in a computer. Each character is designated a decimal number, its ASCII code. As we have already seen, data is stored as a series of 0s and 1s. These binary digits are called *bits*. Standard ASCII uses 8 bits (also defined as 1 byte) to store each character. For example, in ASCII the letter 'a' is denoted by the decimal number 97 which converts to 01100001 in binary. These values are looked up in the standard ASCII table, a small part of which is given at the end of the book. Upper-case letters have different ASCII codes.

Consider the character string 'added' which is shown coded in Figure 7.

Character string	a	d	d	e	d
ASCII	97	100	100	101	100
Binary	01100001	01100100	01100100	01100101	01100100

7. A coded character string.

So 'added' takes 5 bytes or 5 * 8 = 40 bits of storage. Given Figure 7, decoding is accomplished using the ASCII code table. This is not an economical way of encoding and storing data; 8 bits per character seems excessive and no account is taken of the fact that in text documents some letters are used much more frequently than others. There are many lossless data compression models, such as the Huffman algorithm, which uses less storage space by variable length encoding, a technique based on how often a particular letter occurs. Those letters with the highest occurrence are given shorter codes.

Taking the string 'added' again we note that 'a' occurs once, 'e' occurs once, and 'd' occurs three times. Since 'd' occurs most frequently, it should be assigned the shortest code. To find the Huffman code for each letter we count the letters of 'added' as follows:

$$1a \rightarrow 1e \rightarrow 3d$$

Next, we find the two letters that occur least frequently, namely 'a' and 'e', and we form the structure in Figure 8, called a *binary tree*. The number 2 at the top of the tree is found by adding the number of occurrences of the two least frequent letters.

In Figure 9, we show the new node representing three occurrences of the letter 'd'.

Figure 9 shows the completed tree with total number of letter occurrences at the top. Each edge of the tree is coded as either

8. A binary tree.

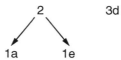

9. The binary tree with a new node.

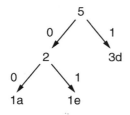

Letter	Code (bits)
a	00
e	10
d	1

10. Completed binary tree.

0 or 1, as shown in Figure 10, and the codes are found by following the paths up the tree.

So 'added' is coded as a = 00, d = 1, d = 1, e = 10, d = 1, which gives us 0011101. Using this method we see that 3 bits are used for storing the letter 'd', 2 bits for letter 'a', and 2 bits for letter 'e', giving a total of 7 bits. This is a big improvement on the original 40 bits.

A way of measuring the efficiency of compression is to use the data compression ratio, which is defined as the uncompressed size of a file divided by its compressed size. In this example, 45/7 is approximately equal to 6.43, a high compression rate, showing good storage savings. In practice these trees are very large and are optimized using sophisticated mathematical techniques. This example has shown how we can compress data without losing any of the information contained in the original file and it is therefore called lossless compression.

Lossy data compression

In comparison, sound and image files are usually much larger than text files and so another technique called *lossy compression* is used. This is because, when we are dealing with sound and images, lossless compression methods may simply not result in a sufficiently high compression ratio for data storage to be viable. Equally, some data loss is tolerable for sound and images. Lossy compression exploits this latter feature by permanently removing some data in the original file so reducing the amount of storage space needed. The basic idea is to remove some of the detail without overly affecting our perception of the image or sound.

For example, consider a black and white photograph, more correctly described as a *greyscale image*, of a child eating an ice-cream at the seaside. Lossy compression removes an equal amount of data from the image of the child and that of the sea. The percentage of data removed is calculated such that it will not have a significant impact on the viewer's perception of the resulting (compressed) image—too much compression will lead to a fuzzy photo. There's a trade-off between the level of compression and quality of picture.

If we want to compress a greyscale image, we first divide it into blocks of 8 pixels by 8 pixels. Since this is a very small area, all the pixels are generally similar in tone. This observation, together with knowledge about how we perceive images, is fundamental to lossy compression. Each pixel has a corresponding numeric value between 0 for pure black and 255 for pure white, with the numbers between representing shades of grey. After some further processing using a method called the Discrete Cosine Algorithm, an average intensity value for each block is found and the results compared with each of the actual values in a given block. Since we are comparing these actual values to the average most of them will be 0, or 0 when rounded. Our lossy algorithm collects all these 0s together, which represent the information from the pixels that is less important to the image. These values, corresponding to high frequencies in our image, are all grouped together and the redundant information is removed, using a technique called *quantization*, resulting in compression. For example if out of sixty-four values each requiring 1 byte of storage, we have twenty 0s, then after compression we need only 45 bytes of storage. This process is repeated for all the blocks that make up the image and so redundant information is removed throughout.

For colour images the JPEG (Joint Photographic Experts Group) algorithm, for example, recognizes red, blue, and green, and assigns each a different weight based on the known properties of human visual perception. Green is weighted greatest since the human eye is more perceptive to green than to red or blue. Each pixel in a colour image is assigned a red, blue, and green weighting, represented as a triple <R,G,B>. For technical reasons, <R,G,B> triples are usually converted into another triple, <YCbCr> where Y represents the intensity of the colour and both Cb and Cr are chrominance values, which describe the actual colour. Using a complex mathematical algorithm it is possible to reduce the values of each pixel and ultimately achieve lossy compression by reducing the number of pixels saved.

Multimedia files in general, because of their size, are compressed using lossy methods. The more compressed the file, the poorer the reproduction quality, but because some of the data is sacrificed, greater compression ratios are achievable, making the file smaller.

Following an international standard for image compression first produced in 1992 by the JPEG, the JPEG file format provides the most popular method for compressing both colour and greyscale photographs. This group is still very active and meets several times a year.

Consider again the example of a black and white photograph of a child eating an ice-cream at the seaside. Ideally, when we compress this image we want the part featuring the child to remain sharp, so in order to achieve this we would be willing to sacrifice some clarity in the background details. A new method, called *data warping compression*, developed by researchers at

Henry Samueli School of Engineering and Applied Science, UCLA, now makes this possible. Those readers interested in the details are referred to the Further reading section at the end of the book.

We have seen how a distributed data file system can be used to store big data. Problems with storage have been overcome to the extent that big data sources can now be used to answer questions that previously we could not answer. As we will see in Chapter 4, an algorithmic method called *MapReduce* is used for processing data stored in the Hadoop DFS.

Chapter 4
Big data analytics

Having discussed how big data is collected and stored, we can now look at some of the techniques used to discover useful information from that data such as customer preferences or how fast an epidemic is spreading. Big data analytics, the catch-all term for these techniques, is changing rapidly as the size of the datasets increases and classical statistics makes room for this new paradigm.

Hadoop, introduced in Chapter 3, provides a means for storing big data through its distributed file system. As an example of big data analytics we'll look at MapReduce, which is a distributed data processing system and forms part of the core functionality of the Hadoop Ecosystem. Amazon, Google, Facebook, and many other organizations use Hadoop to store and process their data.

MapReduce

A popular way of dealing with big data is to divide it up into small chunks and then process each of these individually, which is basically what MapReduce does by spreading the required calculations or queries over many, many computers. It's well worth working through a much simplified and reduced example of how MapReduce works—and as we are doing this by hand it really will need to be a considerably reduced example, but it will still demonstrate the process that would be used for big data. There would be typically many thousands of processors used to process a huge amount of data in parallel, but the process is scalable and it's actually a very ingenious idea and simple to follow.

There are several parts to this analytics model: the *map* component; the *shuffle* step; and the *reduce* component. The map component is written by the user and sorts the data we are interested in. The shuffle step, which is part of the main Hadoop MapReduce code, then groups the data by key, and finally we have the reduce component, which again is provided by the user, which aggregates these groups and produces the result. The result is then sent to HDFS for storage.

For example, suppose we have the following key–value files stored in the Hadoop distributed file system, with statistics on each of the following: measles, Zika virus, TB, and Ebola. The disease is the key and a value representing the number of cases for each disease is given. We are interested in the total number of cases of each disease.

File 1:
Measles,3
Zika,2 TB,1 Measles,1
Zika,3 Ebola,2

File 2:
Measles,4
Zika,2 TB,1

File 3:
Measles,3 Zika,2
Measles,4 Zika,1 Ebola,3

The mapper enables us to read each of these input files separately, line by line, as shown in Figure 11. The mapper then returns the key–value pairs for each of these distinct lines.

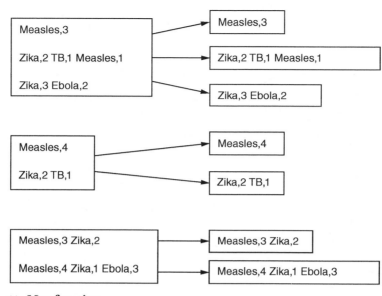

11. Map function.

Having split the files and found key–values for each split, the next step in the algorithm is provided by the master program, which sorts and shuffles the key–values. The diseases are sorted alphabetically and the result is sent to an appropriate file ready for the reducer, as shown in Figure 12.

Continuing to follow Figure 12, the reduce component combines the results of the map and shuffle stages, and as a result sends each disease to a separate file. The reduce step in the algorithm then allows the individual totals to be calculated and these results are sent to a final output file, as key–value pairs, which can be saved in the DFS.

This is a very small example, but the MapReduce method enables us to analyse very large amounts of data. For example, using the data supplied by Common Crawl, a non-profit organization that

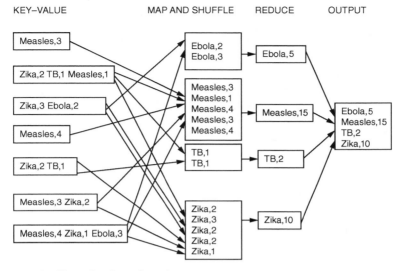

| KEY–VALUE | MAP AND SHUFFLE | REDUCE | OUTPUT |

12. Shuffle and reduce functions.

provides a free copy of the Internet, we could count the number of times each word occurs on the Internet by writing a suitable computer program that uses MapReduce.

Bloom filters

A particularly useful method for mining big data is the Bloom filter, a technique based on probability theory which was developed in the 1970s. As we will see, Bloom filters are particularly suited to applications where storage is an issue and where the data can be thought of as a list.

The basic idea behind Bloom filters is that we want to build a system, based on a list of data elements, to answer the question 'Is X in the list?' With big datasets, searching through the entire set may be too slow to be useful, so we use a Bloom filter which, being a probabilistic method, is not 100 per cent accurate—the algorithm may decide that an element belongs to the list when actually it does not; but it is a fast, reliable, and storage efficient method of extracting useful knowledge from data.

Bloom filters have many applications. For example, they can be used to check whether a particular Web address leads to a malicious website. In this case, the Bloom filter would act as a blacklist of known malicious URLs against which it is possible to check, quickly and accurately, whether it is likely that the one you have just clicked on is safe or not. Web addresses newly found to be malicious can be added to the blacklist. Since there are now over a billion websites, and more being added daily, keeping track of malicious sites is a big data problem.

A related example is that of malicious email messages, which may be spam or may contain phishing attempts. A Bloom filter provides us with a quick way of checking each email address and hence we would be able to issue a timely warning if appropriate. Each address occupies approximately 20 bytes, so storing and checking each of them becomes prohibitively time-consuming since we need to do this very quickly—by using a Bloom filter we are able to reduce the amount of stored data dramatically. We can see how this works by following the process of building a small Bloom filter and showing how it would function.

Suppose we have the following list of email addresses that we want to flag as malicious: <aaa@aaaa.com>; <bbb@nnnn.com>; <ccc@ ff.com>; <dd@ggg.com>. To build our Bloom filter first assume we have 10 bits of memory available on a computer. This is called a *bit array* and initially it is empty. A bit has just two states, usually denoted by 0 and 1, so we will start by setting all values in the bit array to 0, meaning empty. As we will see shortly, a bit with a value of 1 will mean the associated index has been assigned at least once.

The size of our bit array is fixed and will remain the same regardless of how many cases we add. We index each bit in the array as shown in Figure 13.

Index	0	1	2	3	4	5	6	7	8	9
Bit value	0	0	0	0	0	0	0	0	0	0

13. 10-bit array.

We now need to introduce *hash functions*, which are algorithms designed to map each element in a given list to one of the positions in the array. By doing this, we now store only the mapped position in the array, rather than the email address itself, so that the amount of storage space required is reduced.

For our demonstration, we show the result of using two hash functions, but typically seventeen or eighteen functions would be used together with a much bigger array. Since these functions are designed to map more or less uniformly, each index has an equal chance of being the result each time the hash algorithm is applied to a different address.

So, first we let the hash algorithms assign each email address to one of the indices of the array.

To add 'aaa@aaaa.com' to the array, it is first passed through hash function 1, which returns an array index or position value. For example, let's say hash function 1 returned index 3. Hash function 2, again applied to 'aaa@aaaa.com', returned index 4. These two positions will each have their stored bit value set to 1. If the position was already set to 1 then it would be left alone. Similarly, adding 'bbb@nnnn.com' may result in positions 2 and 7 being occupied or set to 1 and 'ccc@ff.com' may return positions 4 and 7. Finally, assume the hash functions applied to 'dd@ggg.com' return the positions 2 and 6. These results are summarized in Figure 14.

The actual Bloom filter array is shown in Figure 15 with occupied positions having a value set to 1.

DATA	HASH 1	HASH 2
aaa@aaaa.com	3	4
bbb@nnnn.com	2	7
ccc@ff.com	4	7
dd@ggg.com	2	6

14. Summary of hash function results.

Index	0	1	2	3	4	5	6	7	8	9
Bit value	0	0	1	1	1	0	1	1	0	0

15. Bloom filter for malicious email addresses.

So, how do we use this array as a Bloom filter? Suppose, now, that we receive an email and we wish to check whether the address appears on the malicious email address list. Suppose it maps to positions 2 and 7, both of which have value 1. Because all values returned are equal to 1 it *probably* belongs to the list and so is *probably* malicious. We cannot say for certain that it belongs to the list because positions 2 and 7 have been the result of mapping other addresses and indexes may be used more than once. So the result of testing an element for list membership also includes the probability of returning a false positive. However, if an array index with value 0 is returned by any hash function (and, remember, there would generally be seventeen or eighteen functions) we would then definitely know that the address was not on the list.

The mathematics involved is complex but we can see that the bigger the array the more unoccupied spaces there will be and the less chance of a false positive result or incorrect matching. Obviously the size of the array will be determined by the number of keys and hash functions used, but it must be big enough to allow a sufficient number of unoccupied spaces for the filter to function effectively and minimize the number of false positives.

Bloom filters are fast and they can provide a very useful way of detecting fraudulent credit card transactions. The filter checks to see whether or not a particular item belongs to a given list or set, so an unusual transaction would be flagged as not belonging to the list of your usual transactions. For example if you have never purchased mountaineering equipment on your credit card, a Bloom filter will flag the purchase of a climbing rope as suspicious. On the other hand, if you do buy mountaineering equipment, the Bloom filter will identify this purchase as probably acceptable but there will be a probability that the result is actually false.

Bloom filters can also be used for filtering email for spam. Spam filters provide a good example since we do not know exactly what we are looking for—often we are looking for patterns, so if we want email messages containing the word 'mouse' to be treated as spam we also want variations like 'm0use' and 'mou$e' to be treated as spam. In fact, we want all possible, identifiable variations of the word to be identified as spam. It is much easier to filter everything that does not match with a given word, so we would only allow 'mouse' to pass through the filter.

Bloom filters are also used to speed up the algorithms used for Web query rankings, a topic of considerable interest to those who have websites to promote.

PageRank

When we search on Google, the websites returned are ranked according to their relevance to the search terms. Google achieves this ordering primarily by applying an algorithm called PageRank. The name PageRank is popularly believed to have been chosen after Larry Page, one of the founders of Google, who, working with co-founder Sergey Brin, published articles on this new algorithm. Until the summer of 2016, PageRank results were publicly available by downloading the Toolbar PageRank. The public PageRank tool was based on a range from 1 and 10. Before it was withdrawn, I saved a few results. If I typed 'Big Data' into Google using my laptop, I got a message informing me there were 'About 370,000,000 results (0.44 seconds)' with a PageRank of 9. Top of this list were some paid advertisements, followed by Wikipedia. Searching on 'data' returned about 5,530,000,000 results in 0.43 seconds with a PageRank of 9. Other examples, all with a PageRank of 10, included the USA government website, Facebook, Twitter, and the European University Association.

This method of calculating a PageRank is based on the number of links pointing to a webpage—the more links, the higher the score, and the higher the page appears as a search result. It does not reflect the number of times a page is visited. If you are a website designer, you want to optimize your website so that it appears very near the top of the list given certain search terms, since most people do not look further than the first three or four results. This requires a huge number of links and as a result, almost inevitably, a trade in links became established. Google tried to address this 'artificial' ranking by assigning a new ranking of 0 to implicated companies or even by removing them completely from Google, but this did not solve the problem; the trade was merely forced underground, and links continued to be sold.

PageRank itself has not been abandoned and forms part of a large suite of ranking programs which are not available for public viewing. Google re-calculates rankings regularly in order to reflect added links as well as new websites. Since PageRank is commercially sensitive, full details are not publicly available but we can get the general idea by looking at an example. The algorithm provides a complex way of analysing the links between webpages based on probability theory, where probability 1 indicates certainty and probability 0 indicates impossibility, with everything else having a probability value somewhere in-between.

Big Data

To understand how the ranking works, we first need to know what a probability distribution looks like. If we think of the result of rolling a fair six-sided die, each of the outcomes 1 through 6 is equally likely to occur and so each has a probability of 1/6. The list of all the possible outcomes, together with the probability associated with each, describes a probability distribution.

Going back to our problem of ranking webpages according to importance, we cannot say that each is equally important, but if we had a way of assigning probabilities to each webpage, this would give us a reasonable indication of importance. So what algorithms such as PageRank do is construct a probability distribution for the entire Web. To explain this, let's consider a random surfer of the Web, who starts at any webpage and then moves to another page using the links available.

We will consider a simplified example where we have a web consisting of only three webpages; BigData1, BigData2, and BigData3. Suppose the only links are from BigData2 to BigData3, BigData2 to BigData1, and BigData1 to BigData3. Then our web can be represented as shown in Figure 16, where the nodes are webpages and the arrows (edges) are links.

Each page has a PageRank indicating its importance or popularity. BigData3 will be the most highly ranked because it has the most links going to it, making it the most popular. Suppose now that

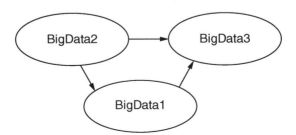

16. **Directed graph representing a small part of the Web.**

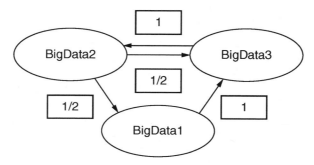

17. Directed graph representing a small part of the Web with added link.

a random surfer visits a webpage, he or she has one proportional vote to cast, which is divided equally between the next choices of webpage. For example, if our random surfer is currently visiting BigData1, the only choice is to then visit BigData3. So we can say that a vote of 1 is cast for BigData3 by BigData1.

In the real Web new links are made all the time, so suppose we now find that BigData3 links to BigData2, as shown in Figure 17, then the PageRank for BigData2 will have changed because the random surfer now has a choice of where to go after BigData3.

If our random surfer starts off at BigData1, then the only choice is to visit BigData3 next and so the total vote of 1 goes to BigData3, and BigData2 gets a vote of 0. If he or she starts at BigData2 the vote is split equally between BigData3 and BigData1. Finally, if the random surfer starts at BigData3 his or her entire vote is cast for BigData2. These proportional 'votes' are summarized in the Figure 18.

Using Figure 18, we now see the total votes cast for each webpage as follows:

Total votes for BD1 are 1/2 (coming from BD2)
Total votes for BD2 are 1 (coming from BD3)
Total votes for BD3 are 1½ (coming from BD 1 and BD2)

	Proportion of vote cast by BD1	Proportion of vote cast by BD2	Proportion of vote cast by BD3
For BD1	0	1/2	0
For BD2	0	0	1
For BD3	1	1/2	0

18. Votes cast for each webpage.

Since the choice of starting page for the surfer is random, each one is equally likely and so is assigned an initial PageRank of 1/3. To form the desired PageRanks for our example, we need to update the initial PageRanks according to the proportion of votes cast for each page.

For example, BD1 has just 1/2 vote, cast by BD2, so the PageRank of BD1 is 1/3 * 1/2 = 1/6. Similarly, PageRank BD2 is given by 1/3 * 1 = 2/6 and PageRank BD3 is 1/3 * 3/2 = 3/6. Since all the Page Rankings now add up to one, we have a probability distribution which shows the importance, or rank, of each page.

But there is a complication here. We said that the probability that a random surfer was on any page initially was 1/3. After one step, we have calculated the probability that a random surfer is on BD1 is 1/6. What about after two steps? Well, again we use the current PageRanks as votes to calculate the new PageRanks. The calculations are slightly different for this round because the current PageRanks are not equal but the method is the same, giving new PageRanks as follows: PageRank BD1 is 2/12, PageRank BD2 is 6/12, and PageRank BD3 is 4/12. These steps, or iterations, are repeated until the algorithm converges, meaning that the process continues like this until no more changes can be made by any further multiplication. Having achieved a final ranking, PageRank can select the page with the highest ranking for a given search.

Page and Brin, in their original research papers, presented an equation for calculating the PageRank which included a Damping Factor d, defined as the probability that a random Web surfer will click on one of the links on the current page. The probability that a random Web surfer will *not* click on one of the links on the current page is therefore $(1 - d)$, meaning that the random surfer has finished surfing. It was this Damping Factor that ensured the PageRank averaged over the entire Web settles down to 1, after a sufficient number of iterative calculations. Page and Brin reported that a web consisting of 322 million links settled down after fifty-two iterations.

Public datasets

There are many freely available big datasets that interested groups or individuals can use for their own projects. Common Crawl, mentioned earlier in this chapter, is one example. Hosted by the Amazon Public Datasets Program, in October 2016 the Common Crawl monthly archive contained more than 3.25 billion webpages. Public datasets are in a broad range of specialties, including genome data, satellite imagery, and worldwide news data. For those not likely to write their own code, Google's Ngram Viewer provides an interesting way of exploring some big datasets interactively (see Further reading for details).

Big data paradigm

We have seen some of the ways in which big data can be useful and in Chapter 2 we talked about small data. For small data analysis, the scientific method is well-established and necessarily involves human interaction: someone comes up with an idea, formulates a hypothesis or model, and devises ways to test its predictions. Eminent statistician George Box wrote in 1978, 'all models are wrong, but some are useful'. The point he makes is that statistical and scientific models in general do not provide exact representations of the world about us, but a good model can provide a useful picture on which to base predictions and draw conclusions confidently. However, as we have shown, when working with big data we do not follow this method. Instead we find that the machine, not the scientist, is predominant.

Writing in 1962, Thomas Kuhn described the concept of scientific revolutions, which follow long periods of normal science when an existing paradigm is developed and investigated to the full. If sufficiently intractable anomalies occur to undermine the existing theory, resulting in loss of confidence by researchers, then this is termed a 'crisis', and it is ultimately resolved by a new theory or paradigm. For a new paradigm to be accepted, it must answer some of the questions found to be problematic in the old paradigm. However, in general, a new paradigm does not completely overwhelm the previous one. For example, the shift from Newtonian mechanics to Einstein's relativity theory changed the way science viewed the world, without making Newton's laws obsolete: Newtonian mechanics now form a special case of the wider ranging relativity theory. Shifting from classical statistics to big data analytics also represents a significant change, and has many of the hallmarks of a paradigm shift. So techniques will inevitably need to be developed to deal with this new situation.

Consider the technique of finding correlations in data, which provides a means of prediction based on the strength of the relationships between variables. It is acknowledged in classical statistics that correlation does not imply causation. For example, a teacher may document both the number of student absences from lectures and student grades; and then, on finding an apparent correlation, he or she may use absences to predict grades. However, it would be incorrect to conclude that absences cause poor grades. We cannot know why the two variables are correlated just by looking at the blind calculations: maybe the less able students tend to miss class; maybe students who are absent due to sickness cannot later catch up. Human interaction and interpretation is needed in order to decide which correlations are useful.

With big data, using correlation creates additional problems. If we consider a massive dataset, algorithms can be written that, when applied, return a large number of spurious correlations that are totally independent of the views, opinions, or hypotheses of any human being. Problems arise with false correlations—for example, divorce rate and margarine consumption, which is just one of many spurious correlations reported in the media. We can see the absurdity of this correlation by applying scientific method. However, when the number of variables becomes large, the number of spurious correlations also increases. This is one of the main problems associated with trying to extract useful information from big data, because in doing so, as with mining big data, we are usually looking for patterns and correlations. As we will see in Chapter 5, one of the reasons Google Flu Trends failed in its predictions was because of these problems.

Chapter 5
Big data and medicine

Big data analysis is significantly changing the world of healthcare. Its potential has yet to be fully realized but includes medical diagnosis, epidemic prediction, gauging the public response to government health warnings, and the reduction in costs associated with healthcare systems. Let's start by looking at what is now termed *healthcare informatics*.

Healthcare informatics

Medical big data is collected, stored, and analysed using the general techniques described in previous chapters. Broadly speaking, healthcare informatics and its many sub-disciplines, such as clinical informatics and bio-informatics, use big data to provide improved patient care and reduce costs. Consider the definition criteria for big data (discussed in Chapter 2)—volume, variety, velocity, and veracity—and how they apply to medical data. Volume and velocity are satisfied, for example, when public-health-related data is collected through social networking sites for epidemic tracking; variety is satisfied since patient records are stored in text format, both structured and unstructured, and sensor data such as that provide by MRIs is also collected; veracity is fundamental to medical applications and considerable care is taken to eliminate inaccurate data.

Social media is a potentially valuable source of medically related information through data collection from sites such as Facebook, Twitter, various blogs, message boards, and Internet searches. Message boards focused on specific healthcare issues are abundant, providing a wealth of unstructured data. Posts on both Facebook and Twitter have been mined, using classification techniques similar to those described in Chapter 4, to monitor experience of unpleasant reactions to medications and supply healthcare professionals with worthwhile information regarding drug interactions and drug abuse. Mining social media data for public-health-related research is now a recognized practice within the academic community.

Designated social networking sites for medical professionals, such as Sermo Intelligence, a worldwide medical network and self-proclaimed 'largest global healthcare data collection company', provide healthcare personnel with instant crowdsourcing benefits from interaction with their peers. Online medical advice sites are becoming increasingly popular and generate yet more information. But, although not publicly accessible, perhaps the most important source is the vast collection of Electronic Health Records. These records, usually referred to simply by their initials, EHR, provide an electronic version of a patient's full medical history, including diagnoses, medications prescribed, medical images such as X-rays, and all other relevant information collected over time, thus constructing a 'virtual patient'—a concept we will look at later in this chapter. As well as using big data to improve patient care and cut costs, by pooling the information generated from a variety of online sources it becomes possible to think in terms of predicting the course of emerging epidemics.

Google Flu Trends

Every year, like many countries, the US experiences an influenza (or flu) epidemic resulting in stretched medical resources and considerable loss of life. Data from past epidemics supplied by the US Center for Disease Control (CDC), the public health monitoring agency, together with big data analytics, provide the driving force behind researchers' efforts to predict the spread of the illness in order to focus services and reduce its impact.

The Google Flu Trends team started working on predicting flu epidemics using search engine data. They were interested in how the course of the annual flu epidemic might be predicted faster than it currently took the CDC to process its own data. In a letter published in the prestigious scientific journal *Nature* in February 2009, the team of six Google software engineers explained what they were doing. If data could be used to accurately predict the course of the annual US flu epidemic then the illness could be contained, saving lives and medical resources. The Google team explored the idea that this could be achieved by collecting and analysing search engine queries relevant to concerns about the flu. Previous attempts to use online data to predict the spread of the flu had either failed or been met with limited success. However, by learning from earlier mistakes in this pioneering research, Google and the CDC hoped to be successful in using big data generated by search engine queries to improve epidemic tracking.

The CDC and its European counterpart, the European Influenza Surveillance Scheme (EISS), collect data from various sources, including physicians, who report on the number of patients they see with flu-like symptoms. By the time this data is collated it is typically about two weeks old and the epidemic has progressed further. Using data collected in real-time from the Internet, the Google/CDC team aimed to improve the accuracy of epidemic predictions and to deliver results within a single day. To do this, data was collected on flu-related search queries varying from individual Internet searches on flu remedies and symptoms to mass data such as phone calls made to medical advice centres. Google was able to tap into a vast amount of search query data that it had accumulated between 2003 and 2008, and by using

IP addresses it was able to identify the geographic location of where search queries had been generated and thus group the data according to State. The CDC data is collected from ten regions, each containing the cumulative data from a group of States (e.g. Region 9 includes Arizona, California, Hawaii, and Nevada), and this was then integrated into the model.

The Google Flu Trends project hinged on the known result that there is a high correlation between the number of flu-related online searches and visits to the doctor's surgery. If a lot of people in a particular area are searching for flu-related information online, it might then be possible to predict the spread of flu cases to adjoining areas. Since the interest is in finding trends, the data can be anonymized and hence no consent from individuals is required. Using their five-year accumulation of data, which they limited to the same time-frame as the CDC data, and so collected only during the flu season, Google counted the weekly occurrence of each of the fifty million most common search queries covering all subjects. These search query counts were then compared with the CDC flu data, and those with the highest correlation were used in the flu trends model. Google chose to use the top forty-five flu-related search terms and subsequently tracked these in the search queries people were making. The complete list of search terms is secret but includes, for example, 'influenza complication', 'cold/flu remedy', and 'general influenza symptoms'. The historical data provided a baseline from which to assess current flu activity on the chosen search terms and by comparing the new real-time data against this, a classification on a scale from 1 to 5, where 5 signified the most severe, was established.

Used in the 2011–12 and 2012–13 US flu seasons, Google's big data algorithm famously failed to deliver. After the flu season ended, its predictions were checked against the CDC's actual data. In building the model, which should be a good representation of flu trends from the data available, the Google Flu Trends algorithm over-predicted the number of flu cases by at least

50 per cent during the years it was used. There were several reasons why the model did not work well. Some search terms were intentionally excluded because they did not fit the expectations of the research team. The much publicized example is that high-school basketball, seemingly unrelated to the flu, was nevertheless highly correlated with the CDC data, but it was excluded from the model. Variable selection, the process by which the most appropriate predictors are chosen, always presents a challenging problem and so is done algorithmically to avoid bias. Google kept the details of their algorithm confidential, noting only that high-school basketball came in the top 100 and justifying its exclusion by pointing out that the flu and basketball both peak at the same time of year.

As we have noted, in constructing their model Google used forty-five search terms as predictors of the flu. Had they only used one, for example 'influenza' or 'flu', important and relevant information such as all the searches on 'cold remedy' would have gone unnoticed and unreported. Accuracy in prediction is improved by having a sufficient number of search terms but it can also decrease if there are too many. Current data is used as training data to construct a model that predicts future data trends, and when there are too many predictors, small random cases in the training data are modelled and so, although the model fits the training data very well, it does not predict well. This seemingly paradoxical phenomenon, called 'over-fitting', was not taken into account sufficiently by the team. Omitting high-school basketball as simply being coincidental to the flu season made sense, but there were fifty million distinct search terms and with such a big number it is almost inevitable that others will correlate strongly with the CDC but not be relevant to flu trends.

Visits to the doctor with flu-like symptoms often resulted in a diagnosis that was not the flu (e.g. it was the common cold). The data Google used, collected selectively from search engine queries, produced results that are not scientifically sound given the obvious bias produced, for example by eliminating everyone who does not use a computer and everyone using other search engines. Another issue that may have led to poor results was that customers searching Google on 'flu symptoms' would probably have explored a number of flu-related websites, resulting in their being counted several times and thus inflating the numbers. In addition, search behaviour changes over time, especially during an epidemic, and this should be taken into account by updating the model regularly. Once errors in prediction start to occur, they tend to cascade, which is what happened with the Google Flu Trends predictions: one week's errors were passed along to the next week. Search queries were considered as they had actually occurred and not grouped according to spelling or phrasing. Google's own example was that 'indications of flu', 'flu indications', and 'indications of the flu' were each counted separately.

The work, which dates back to 2007–8, has been much criticized, sometimes unfairly, but the criticism has usually related to lack of transparency, for example the refusal to reveal all the chosen search terms and unwillingness to respond to requests from the academic community for information. Search engine query data is not the product of a designed statistical experiment and finding a way to meaningfully analyse such data and extract useful knowledge is a new and challenging field that would benefit from collaboration. For the 2012–13 flu season, Google made significant changes to its algorithms and started to use a relatively new mathematical technique called Elasticnet, which provides a rigorous means of selecting and reducing the number of predictors required. In 2011, Google launched a similar program for tracking Dengue fever, but they are no longer publishing predictions and, in 2015, Google Flu Trends was withdrawn. They are, however, now sharing their data with academic researchers.

Google Flu Trends, one of the earlier attempts at using big data for epidemic prediction, provided useful insights to researchers who came after them. Even though the results did not live up to expectations, it seems likely that in the future better techniques will be developed and the full potential of big data in tracking epidemics realized. One such attempt was made by a group of scientists from the Los Alamos National Laboratory in the USA, using data from Wikipedia. The Delphi Research Group at Carnegie Mellon University won the CDC's challenge to 'Predict the Flu' in both 2014–15 and 2015–16 for the most accurate forecasters. The group successfully used data from Google, Twitter, and Wikipedia for monitoring flu outbreaks.

The West Africa Ebola outbreak

The world has experienced many pandemics in the past; the Spanish flu of 1918–19 killed somewhere between twenty million and fifty million people and in total infected about 500 million people. Very little was known about the virus, there was no effective treatment, and the public health response was limited—necessarily so, due to lack of knowledge. This changed in 1948 with the inauguration of the World Health Organization (WHO), charged with monitoring and improving global health through worldwide cooperation and collaboration. On 8 August 2014, at a teleconference meeting of the International Health Regulations Emergency Committee, the WHO announced that an outbreak of the Ebola virus in West Africa formally constituted a 'public health emergency of international concern' (PHEIC). Using a term defined by the WHO, the Ebola outbreak constituted an 'extraordinary event' requiring an international effort of unprecedented proportions in order to contain it and thus avert a pandemic.

The West Africa Ebola outbreak in 2014, primarily confined to Guinea, Sierra Leone, and Liberia, presented a different set of problems to the annual US flu outbreak. Historical data on Ebola was either not available or of little use since an outbreak of these proportions had never been recorded, and so new strategies for dealing with it needed to be developed. Given that knowledge of population movements help public health professionals monitor the spread of epidemics, it was believed that the information held by mobile phone companies could be used to track travel in the infected areas, and measures put in place, such as travel restrictions, that would contain the virus, ultimately saving lives. The resulting real-time model of the outbreak would predict where the next instances of the disease were most likely to occur and resources could be focused accordingly.

The digital information that can be garnered from mobile phones is fairly basic: the phone number of both the caller and the person being called, and an approximate location of the caller—a call made on a mobile phone generates a trail that can be used to estimate the caller's location according to the tower used for each call. Getting access to this data posed a number of problems: privacy issues were a genuine concern as individuals who had not given consent for their calls to be tracked could be identified.

In the West African countries affected by Ebola, mobile phone density was not uniform, with the lowest percentages occurring in poor rural areas. For example, in 2013 just over half the households in Liberia and Sierra Leone, two of the countries directly affected by the outbreak in 2014, had a mobile phone, but even so they could provide sufficient data to usefully track movement.

Some historic mobile phone data was given to the Flowminder Foundation, a non-profit organization based in Sweden, dedicated to working with big data on public health issues that affect the world's poorer countries. In 2008, Flowminder were the first to use mobile operator data to track population movements in a medically challenging environment, as part of an initiative by the WHO to eradicate malaria, so they were an obvious choice to work on the Ebola crisis. A distinguished international team used anonymized historic data to construct maps of population movements in the areas affected by Ebola. This historic data was of limited use since behaviour changes during epidemics, but it does give strong indications of where people will tend to travel, given an emergency. Mobile phone mast activity records provide real-time population activity details.

However, the Ebola prediction figures published by WHO were over 50 per cent higher than the cases actually recorded.

The problems with both the Google Flu Trends and Ebola analyses were similar in that the prediction algorithms used were based only on initial data and did not take into account changing conditions. Essentially, each of these models assumed that the number of cases would continue to grow at the same rate in the future as they had before the medical intervention began. Clearly, medical and public health measures could be expected to have positive effects and these had not been integrated into the model.

The Zika virus, transmitted by Aedes mosquitoes, was first recorded in 1947 in Uganda, and has since spread as far afield as Asia and the Americas. The current Zika virus outbreak, identified in Brazil in 2015, resulted in another PHEIC. Lessons have been learned regarding statistical modelling with big data from work by Google Flu Trends and during the Ebola outbreak, and it is now generally acknowledged that data should be collected from multiple sources. Recall that the Google Flu Trends project collected data only from its own search engine.

The Nepal earthquake

So what is the future for epidemic tracking using big data? The real-time characteristics of mobile phone call detail records (CDRs) have been used to assist in monitoring population movements in disasters as far ranging as the Nepal earthquake and the swine-flu outbreak in Mexico. For example, an international Flowminder team, with scientists from the Universities of Southampton and Oxford, as well as institutions in the US and

China, following the Nepal earthquake of 25 April 2015, used CDRs to provide estimates of population movements. A high percentage of the Nepali population has a mobile phone and by using the anonymized data of twelve million subscribers, the Flowminder team was able to track population movements within nine days of the earthquake. This quick response was due in part to having in place an agreement with the main service provider in Nepal, technical details of which were only completed a week before the disaster. Having a dedicated server with a 20 Tb hard drive in the providers' data centre enabled the team to start work immediately, resulting in information being made available to disaster relief organizations within nine days of the earthquake.

Big data and smart medicine

Every time a patient visits a doctor's office or hospital, electronic data is routinely collected. Electronic health records constitute legal documentation of a patient's healthcare contacts: details such as patient history, medications prescribed, and test results are recorded. Electronic health records may also include sensor data such as Magnetic Resonance Imaging (MRI) scans. The data may be anonymized and pooled for research purposes. It is estimated that in 2015, an average hospital in the USA will store over 600 Tb of data, most of which is unstructured. How can this data be mined to give information that will improve patient care and cut costs? In short, we take the data, both structured and unstructured, identify features relevant to a patient or patients, and use statistical techniques such as classification and regression to model outcomes. Patient notes are primarily in the format of unstructured text, and to effectively analyse these requires natural language processing techniques such as those used by IBM's Watson, which is discussed in the next section.

According to IBM, by 2020 medical data is expected to double every seventy-three days. Increasingly used for monitoring healthy individuals, wearable devices are widely used to count the number of steps we take each day; measure and balance our calorie requirements; track our sleep patterns; as well as giving immediate information on our heart rate and blood pressure. The information gleaned can then be uploaded onto our PCs and records kept privately or, as is sometimes the case, shared voluntarily with employers. This veritable cascade of data on individuals will provide healthcare professionals with valuable public health data as well as providing a means for recognizing changes in individuals that might help avoid, for example, a heart attack. Data on populations will enable physicians to track, for example, side-effects of a particular medication based on patient characteristics.

Following the completion of the Human Genome Project in 2003, genetic data will increasingly become an important part of our individual medical records as well as providing a wealth of research data. The aim of the Human Genome Project was to map all the genes of humans. Collectively, the genetic information of an organism is called its genome. Typically, the human genome contains about 20,000 genes and mapping such a genome requires about 100 Gb of data. Of course, this is a highly complex, specialized, and multi-faceted area of genetic research, but the implications following the use of big data analytics are of interest. The information about genes thus collected is kept in large databases and there has been concern recently that these might be hacked and patients who contributed DNA would be identified. It has been suggested that, for security purposes, false information should be added to the database, although not enough to render it useless for medical research. The interdisciplinary field of bioinformatics has flourished as a consequence of the need to manage and analyse the big data generated by genomics. Gene sequencing has become increasingly rapid and much cheaper in recent years, so that mapping individual genomes is now practical. Taking into account the cost of fifteen years of research, the first human genome sequencing cost nearly US$3 million. Many companies now offer genome sequencing services to individuals at an affordable price.

Growing out of the Human Genome Project, the Virtual Physiological Human (VPH) project aims to build computer representations that will allow clinicians to simulate medical treatments and find the best for a given patient, built on the data from a vast data bank of actual patients. By comparing those with similar symptoms and other medically relevant details, the computer model can predict the likely outcome of a treatment on an individual patient. Data mining techniques are also used and potentially merged with the computer simulations to personalize medical treatment, and so the results of an MRI might integrate with a simulation. The digital patient of the future would contain all the information about a real patient, updated according to smart device data. However, as is increasingly the case, data security is a significant challenge faced by the project.

Watson in medicine

In 2007, IBM decided to build a computer to challenge the top competitors in the US television game show, *Jeopardy*. Watson, a big data analytics system named after the founder of IBM, Thomas J. Watson, was pitted against two *Jeopardy* champions: Brad Rutter, with a winning streak of seventy-four appearances; and Ken Jennings, who had won a staggering total of US$3.25 million. *Jeopardy* is a quiz show in which the host of the show gives an 'answer' and the contestant has to guess the 'question'. There are three contestants and the answers or clues come in several categories such as science, sport, and world history together with less standard, curious categories such as 'before and after'. For example, given the clue 'His tombstone in a Hampshire churchyard reads "knight, patriot, physician and man of letters; 22 May 1859–7 July 1930"', the answer is 'Who is Sir Arthur Conan Doyle?'. In the less obvious category 'catch these men', given the clue 'Wanted for 19 murders, this Bostonian went on the run in 1995 and was finally nabbed in Santa Monica in 2011', the answer is 'Who was Whitey Bulger?'. Clues that were delivered to Watson as text and audio-visual cues were omitted from the competition.

Natural language processing (NLP), as it is known in artificial intelligence (AI), represents a huge challenge to computer science and was crucial to the development of Watson. Information also has to be accessible and retrievable, and this is a problem in machine learning. The research team started out by analysing *Jeopardy* clues according to their lexical answer type (LAT), which classifies the kind of answer specified in the clue. For the second of these examples, the LAT is 'this Bostonian'. For the first example, there is no LAT, the pronoun 'it' does not help. Analysing 20,000 clues the IBM team found 2,500 unique LATs but these covered only about half the clues. Next, the clue is parsed to identify key words and the relationships between them. Relevant documents are retrieved and searched from the computer's structured and unstructured data. Hypotheses are generated based on the initial analyses, and by looking for deeper evidence potential answers are found.

To win *Jeopardy*, fast advanced natural language processing techniques, machine learning, and statistical analysis were crucial. Among other factors to consider were accuracy and choice of category. A baseline for acceptable performance was computed using data from previous winners. After several attempts, deep question and answer analysis, or 'DeepQA', an amalgamation of many AI techniques gave the solution. This system uses a large bank of computers, working in parallel but not connected to the Internet; it is based on probability and the evidence of experts. As well as generating an answer, Watson uses confidence-scoring algorithms to enable the best result to be found. Only when the confidence threshold is reached does Watson indicate that it is ready to give an answer, the equivalent of a human contestant hitting their buzzer. Watson beat the two *Jeopardy* champions.

Jennings, generous in defeat, is quoted as saying, 'I, for one, welcome our new computer overlords'.

The Watson medical system, based on the original *Jeopardy* Watson, retrieves and analyses both structured and unstructured data. Since it builds its own knowledge base it is essentially a system that appears to model human thought processes in a particular domain. Medical diagnoses are based on all available medical knowledge, they are evidence-based, accurate to the extent that the input is accurate and contains all the relevant information, and consistent. Human doctors have experience but are fallible and some are better diagnosticians than others. The process is similar to the Watson of *Jeopardy*, taking into account all the relevant information and returning diagnoses, each with a confidence rating. Watson's built-in AI techniques enable the processing of big data, including the vast amounts generated by medical imaging.

The Watson supercomputer is now a multi-application system and a huge commercial success. In addition Watson has been engaged in humanitarian efforts, for example through a specially developed openware analytics system to assist in tracking the spread of Ebola in Sierra Leone.

Medical big data privacy

Big data evidently has potential to predict the spread of disease and to personalize medicine, but what of the other side of the coin—the privacy of the individual's medical data? Particularly with the growing use of wearable devices and smartphone apps, questions arise as to who owns the data, where it is being stored, who can access and use it, and how secure it is from cyber-attacks. Ethical and legal issues are abundant but not addressed here.

Data from a fitness tracker may become available to an employer and used: favourably, for example to offer bonuses to those who meet certain metrics; or, unfavourably, to determine those who fail to reach the required standards, perhaps leading to an unwanted redundancy offer. In September 2016, a collaborative research team of scientists from the Technische Universität Darmstadt in Germany and the University of Padua in Italy, published the results of their study into fitness tracker data security. Alarmingly, out of the seventeen fitness trackers tested, all from different manufacturers, none was sufficiently secure to stop changes being made to the data and only four took any measures, all bypassed by the team's efforts, to preserve data veracity.

In September 2016 following the Rio Olympic Games, from which most Russian athletes were banned following substantiated reports of a state-run doping programme, medical records of top athletes, including the Williams sisters, Simone Byles, and Chris Froome, were hacked and publicly disclosed by a group of Russian cyber-hackers on the website FancyBears.net. These medical records, held by the World Anti-Doping Agency (WADA) on their data management system ADAMS, revealed only therapeutic use exemptions and therefore no wrong-doing by the cyber-bullied athletes. It is likely that the initial ADAMS hack was the result of spear-phishing email accounts. This technique, whereby an email appears to be sent by a senior trusted source within an organization, such as a healthcare provider, to a more junior member of the same organization, is used to illegally acquire sensitive information such as passwords and account numbers through downloaded malware.

Proofing big data medical databases from cyber-attacks and hence ensuring patient privacy is a growing concern. Anonymized personal medical data is for sale legally but even so it is sometimes possible to identify individual patients. In a valuable exercise highlighting the vulnerability of supposedly secure data, Harvard Data Privacy Lab scientists Latanya Sweeney and Ji Su Yoo, using legally available *encrypted* (i.e. scrambled so that they cannot easily be read; see Chapter 7) medical data originating in

South Korea, were able to decrypt unique identifiers within the records, and identify individual patients through cross-checking with public records.

Medical records are extremely valuable to cyber-criminals. In 2015, the health insurer Anthem declared that its databases had been hacked with over seventy million people affected. Data critical to individual identification, such as name, address, and social security number, was breached by Deep Panda, a Chinese hacking group, using a stolen password to access the system and instal Trojan-horse malware. Critically, the social security numbers, a unique identifier in the USA, were not encrypted, leaving wide open the possibility of identity theft. Many security breaches start with human error: people are busy and do not noticed subtle changes in a Uniform Resource Locator (URL); devices such as flash drives are lost, stolen, and even on occasion deliberately planted, with malware instantly installed once an unsuspecting employee plugs the device into a USB port. Both discontented employees and genuine employee mistakes also account for countless data leaks.

New big data incentives in the management of healthcare are being launched at an increasing rate by world-renowned institutions such as the Mayo Clinic and Johns Hopkins Medical in the USA, the UK's National Health Service (NHS), and Clermont-Ferrand University Hospital in France. Cloud-based systems give authorized users access to data anywhere in the world. To take just one example, the NHS plans to make patient records available via smartphone by 2018. These developments will inevitably generate more attacks on the data they employ, and considerable effort will need to be expended in the development of effective security methods to ensure the safety of that data.

Chapter 6
Big data, big business

In the 1920s, J. Lyons and Co., a British catering firm famous for their 'Corner House' cafés, employed a young Cambridge University mathematician, John Simmons, to do statistical work. In 1947, Raymond Thompson and Oliver Standingford, both of whom had been recruited by Simmons, were sent on a fact-finding visit to the USA. It was on this visit that they first became aware of electronic computers and their potential for executing routine calculations. Simmons, impressed by their findings, sought to persuade Lyons to acquire a computer.

Collaboration with Maurice Wilkes, who was then engaged in building the Electronic Delay Storage Automatic Computer (EDSAC) at the University of Cambridge, resulted in the Lyons Electronic Office. This computer ran on punched cards and was first used by Lyons in 1951 for basic accounting tasks, such as adding up columns of figures. By 1954, Lyons had formed its own computer business and was building the LEO II, followed by the LEO III. Although the first office computers were being installed as early as the 1950s, given their use of valves (6,000 in the case of the LEO I) and magnetic tape, and their very small amount of RAM, these early machines were unreliable and their applications were limited. The original Lyons Electronic Office became widely referred to as the first business computer, paving the way for modern e-commerce and, after several mergers, finally became part of the newly formed International Computers Limited (ICL) in 1968.

e-Commerce

The LEO machines and the massive mainframe computers that followed were suitable only for the number-crunching tasks involved in such tasks as accounting and auditing. Workers who had traditionally spent their days tallying columns of figures now spent their time producing punched cards instead, a task no less tedious while requiring the same high degree of accuracy.

Since the use of computers became feasible for commercial enterprises, there has been interest in how they can be used to improve efficiency, cut costs, and generate profits. The development of the transistor and its use in commercially available computers resulted in ever-smaller machines, and in the early 1970s the first personal computers were being introduced. However, it was not until 1981, when International Business Machines (IBM) launched the IBM-PC on the market, with the use of floppy disks for data storage, that the idea really took off for business. The word-processing and spreadsheet capabilities of succeeding generations of PCs were largely responsible for relieving much of the drudgery of routine office work.

The technology that facilitated electronic data storage on floppy disks soon led to the idea that in future, businesses could be run effectively without the use of paper. In 1975 an article in the American magazine *BusinessWeek* speculated that the almost paper-free office would be a reality by 1990. The suggestion was that by eliminating or significantly reducing the use of paper, an office would become more efficient and costs would be reduced. Paper use in offices declined for a while in the 1980s when much of the paperwork that used to be found in filing cabinets was transferred to computers, but it then rose to an all-time high in 2007, with photocopies accounting for the majority of the increase. Since 2007, paper use has been gradually declining, thanks largely to the increased use of mobile smart devices and facilities such as the electronic signature.

Although the optimistic aspiration of the early digital age to make an office paperless has yet to be fulfilled, the office environment has been revolutionized by email, word-processing, and electronic spreadsheets. But it was the widespread adoption of the Internet that made e-commerce a practical proposition.

Online shopping is perhaps the most familiar example. As customers, we enjoy the convenience of shopping at home and avoiding time-consuming queues. The disadvantages to the customer are few but, depending on the type of transaction, the lack of contact with a store employee may inhibit the use of online purchasing. Increasingly, these problems are being overcome by online customer advice facilities such as 'instant chat', online reviews, and star rankings, a huge choice of goods and services together with generous return policies. As well as buying and paying for goods, we can now pay our bills, do our banking, buy airline tickets, and access a host of other services all online.

eBay works rather differently and is worth mentioning because of the huge amounts of data it generates. With transactions being made through sales and auction bids, eBay generates approximately 50 Tb of data a day, collected from every search, sale, and bid made on their website by a claimed 160 million active users in 190 countries. Using this data and the appropriate analytics they have now implemented recommender systems similar to those of Netflix, discussed later in this chapter.

Social networking sites provide businesses with instant feedback on everything from hotels and vacations to clothes, computers, and yoghurt. By using this information, businesses can see what works, how well it works, and what gives rise to complaints, while fixing problems before they get out of control. Even more valuable is the ability to predict what customers want to buy based on previous sales or website activity. Social networking sites such as Facebook and Twitter collect massive amounts of unstructured data that businesses can benefit from commercially given the appropriate analytics. Travel websites, such as TripAdvisor, also share information with third parties.

Pay-per-click advertising

Professionals are now increasingly acknowledging that appropriate use of big data can provide useful information and generate new customers through improved merchandising and use of better targeted advertising. Whenever we use the Web we are almost inevitably aware of online advertising and we may even post free advertisements ourselves on various bidding sites such as eBay.

One of the most popular kinds of advertising follows the pay-per-click model, which is a system by which relevant advertisements pop up when you are doing an online search. If a business wants their advertisement to be displayed in connection with a particular search term, they place a bid with the service provider on a keyword associated with that search term. They also declare a daily maximum budget. The adverts are displayed in order according to a system based in part on which advertiser has bid the highest for that term.

If you click on their advertisement, the advertiser then must pay the service provider what they bid. Businesses only pay when an interested party clicks on their advertisement, so these adverts must be a good match for the search term to make it more likely that a Web surfer will click on them. Sophisticated algorithms ensure that for the service provider, for example Google or Yahoo, revenue is maximized. The best known implementation of pay-per-click advertising is Google's Adwords. When we search on Google the advertisements that automatically appear on the side of the screen are generated by Adwords. The downside is that clicks can be expensive, and there is also a limit on the number of characters you are allowed to use so that your advertisement will not take up too much space.

Click fraud is also a problem. For example, a rival company may click on your advertisement repeatedly in order to use up your daily budget. Or a malicious computer program, called a clickbot, may be used to generate clicks. The victim of this kind of fraud is the advertiser, since the service provider gets paid and no customers are involved. However, since it is in the best interests of providers to ensure security and so protect a lucrative business, considerable research effort is being made to counteract fraud. Probably the simplest method is to keep track of how many clicks are needed on average to generate a purchase. If this suddenly increases or if there are a large number of clicks and virtually no purchases then fraudulent clicking seems likely.

In contrast to pay-per-click arrangements, targeted advertising is based explicitly on each person's online activity record. To see how this works, we'll start by looking more closely at cookies, which I mentioned briefly in Chapter 1.

Cookies

This term first appeared in 1979 when the operating system UNIX ran a program called Fortune Cookie, which delivered random quotes to the users generated from a large database. Cookies come in several forms, all of which originate externally and are used to keep a record of some activity on a website and/or computer. When you visit a website, a message consisting of a small file that is stored on your computer is sent by a Web server to your browser. This message is one example of a cookie, but there are many other kinds, such as those used for user-authentication purposes and those used for third-party tracking.

Targeted advertising

Every click you make on the Internet is being collected and used for targeted advertising.

This user data is sent to third-party advertising networks and stored on your computer as a cookie. When you click on other sites supported by this network, advertisements for products you looked at previously will be displayed on your screen. Using Lightbeam, a free add-on to Mozilla Firefox, you can keep track of which companies are collecting your Internet activity data.

Recommender systems

Recommender systems provide a filtering mechanism by which information is provided to users based on their interests. Other types of recommender systems, not based on the users' interests, show what other customers are looking at in real-time and often these will appear as 'trending'. Netflix, Amazon, and Facebook are examples of businesses that use these systems.

A popular method for deciding what products to recommend to a customer is *collaborative filtering*. Generally speaking, the algorithm uses data collected on individual customers from their previous purchases and searches, and compares this to a large database of what other customers liked and disliked in order to make suitable recommendations for further purchasing. However, a simple comparison does not generally produce good results. Consider the following example.

Suppose an online bookstore sells a cookery book to a customer. It would be easy to subsequently recommend all cookery books, but this is unlikely to be successful in securing further purchases. There are far too many of them, and the customer already knows he or she likes cookery books. What is needed is a way of reducing

	Daily Salads	Pasta Today	Desserts Tomorrow	Wine For All
Smith	bought		bought	
Jones	bought			bought
Brown		bought	bought	bought

19. Books bought by Smith, Jones, and Brown.

the number of books to those that the customer might actually buy. Let's look at three customers, Smith, Jones, and Brown, together with their book purchases (Figure 19).

The question for the recommender system is which book should be recommended to Smith and which recommended to Jones. We want to know if Smith is more likely to buy *Pasta Today* or *Wine for All*.

To do this we need to use a statistic that is often used for comparing sets and is called the *Jaccard index*. This is defined as the number of items the two sets have in common divided by the total number of distinct items in the two sets. The index measures the similarity between the two sets as the proportion they have in common. The Jaccard distance, defined as one minus the Jaccard index, measures the dissimilarity between them.

Looking again at Figure 19, we see that Smith and Jones have one book purchase in common, *Daily Salads*. Between them they have purchased three distinct books, *Daily Salads*, *Desserts Tomorrow*, and *Wine for All*. This gives them a Jaccard index of 1/3 and a Jaccard distance of 2/3. Figure 20 shows the calculation for all the possible pairs of customers.

Smith and Jones have a higher Jaccard index, or similarity score, than Smith and Brown. This means that Smith and Jones are closer in their purchasing habits—so we recommend *Wine for All*

	Number of titles in common	Total number of distinct titles purchased	Jaccard index	Jaccard distance
Smith and Jones	1	3	1/3	2/3
Smith and Brown	1	4	1/4	3/4
Jones and Brown	1	4	1/4	3/4

20. Jaccard index and distance.

to Smith. What should we recommend to Jones? Smith and Jones have a higher Jaccard index than Jones and Brown, so we recommend *Desserts Tomorrow* to Jones.

Now suppose that customers rate purchases on a five-star system. To make use of this information we need to find other customers who gave the same rating to particular books and see what else they bought as well as considering their purchasing history. The star ratings for each purchase are given in Figure 21.

In this example a different calculation, called the *cosine similarity measure*, which takes the star ratings into account, is described. For this calculation, the information given in the Star Ratings table is represented as vectors. The length or magnitude of the vectors is normalized to 1 and plays no further part in the calculations. The direction of the vectors is used as a way of finding how similar the two vectors are and so who has the best star rating. Based on the theory of vector spaces, a value for the cosine similarity between

	Daily Salads	*Pasta Today*	*Desserts Tomorrow*	*Wine For All*
Smith	5		3	
Jones	2			5
Brown		1	4	3

21. Star ratings for purchases.

the two vectors is found. The calculation is rather different to the familiar trigonometry method, but the basic properties still hold with cosines taking values between 0 and 1. For example, if we find that the cosine similarity between two vectors, each representing a person's star ratings, is 1 then the angle between them is 0 since cosine (0) = 1, and so they must coincide and we can conclude that they have identical tastes. The higher the value of the cosine similarity the greater the similarity in taste.

If you want to see the mathematical details, there are references in the Further reading section at the end of this VSI. What is interesting from our perspective is that the cosine similarity between Smith and Jones works out to be 0.350, and between Smith and Brown it is 0.404. This is a reversal of the previous result, indicating that Smith and Brown have tastes closer than those of Smith and Jones. Informally, this can be interpreted as Smith and Brown being closer in their opinion of *Desserts Tomorrow* than Smith and Jones were in their opinion of *Daily Salads*.

Netflix and Amazon, which we will look at in the next section, both use collaborative filtering algorithms.

Amazon

In 1994, Jeff Bezos founded Cadabra, but soon changed the name to Amazon and in 1995 Amazon.com was launched. Originally an online book store, it is now an international e-commerce company with over 304 million customers worldwide. It produces and sells a diverse range from electronic devices to books and even fresh food items such as yoghurt, milk, and eggs through Amazon Fresh. It is also a leading big data company, with Amazon Web Services providing Cloud-based big data solutions for business, using developments based on Hadoop.

Amazon collected data on what books were bought, what books a customer looked at but did not buy, how long they spent searching, how long they spent looking at a particular book, and whether or not the books they saved were translated into purchases. From this they could determine how much a customer spent on books monthly or annually, and determine whether they were regular customers. In the early days, the data Amazon collected was analysed using standard statistical techniques. Samples were taken of a person and, based on the similarities found, Amazon would offer customers more of the same. Taking this a step further, in 2001 researchers at Amazon applied for and were granted a patent on a technique called item-to-item collaborative filtering. This method finds similar items, not similar customers.

Amazon collects vast amounts of data including addresses, payment information, and details of everything an individual has ever looked at or bought from them. Amazon uses its data in order to encourage the customer to spend more money with them by trying to do as much of the customer's market research as possible. In the case of books, for example, Amazon needs to provide not only a huge selection but to focus recommendations on the individual customer. If you subscribe to Amazon Prime, they also track your movie watching and reading habits. Many customers use smartphones with GPS capability, allowing Amazon to collect data showing time and location. This substantial amount of data is used to construct customer profiles allowing similar individuals and their recommendations to be matched.

Since 2013, Amazon has been selling customer metadata to advertisers in order to promote their Web services operation, resulting in huge growth. For Amazon Web Services, their Cloud computing platform, security is paramount and multi-faceted. Passwords, key-pairs, and digital signatures are just a few of the security techniques in place to ensure that clients' accounts are available only to those with the correct authorization.

Amazon's own data is similarly multi-protected and encrypted, using the AES (Advanced Encryption Standard) algorithm, for storage in dedicated data centres around the world, and Secure Socket Layer (SSL), the industry standard, is used for establishing a secure connection between two machines, such as a link between your home computer and Amazon.com.

Amazon is pioneering *anticipatory shipping* based on big data analytics. The idea is to use big data to anticipate what a customer will order. Initially the idea is to ship a product to a delivery hub before an order actually materializes. As a simple extension, a product can be shipped with a delighted customer receiving a free surprise package. Given Amazon's returns policy, this is not a bad idea. It is anticipated that most customers will keep the items they do order since they are based on their personal preferences, found by using big data analytics. Amazon's 2014 patent on anticipatory shipping also states that goodwill can be bought by sending a promotional gift. Goodwill, increased sales through targeted marketing, and reduced delivery times all make this what Amazon believes to be a worthwhile venture. Amazon also filed for a patent on autonomous flying drone delivery, called Prime Air. In September 2016, the US Federal Aviation Administration relaxed the rules for flying drones by commercial organizations, allowing them, in certain highly controlled situations, to fly beyond the line of sight of the operator. This could be the first stepping stone in Amazon's quest to deliver packages within thirty minutes of an order being placed, perhaps leading to a drone delivery of milk after your smart refrigerator sensor has indicated that you are running out.

Amazon Go, located in Seattle, is a convenience food store and is the first of its kind with no checkout required. As of December 2016 it was only open to Amazon employees and plans for it to be available to the general public in January 2017 have been postponed. At present, the only technical details available are from the patent submitted two years ago, which describes a system eliminating the need to go through an item-by-item checkout. Instead, the details of a customer's actual cart are automatically added to their virtual cart as they shop. Payment is made electronically as they leave the store through a transition area, as long as they have an Amazon account and a smartphone with the Amazon Go app. The Go system is based on a series of sensors, a great many of them, used to identify when an item is taken from or returned to a shelf.

This will generate a huge amount of commercially useful data for Amazon. Clearly, since every shopping action made between entering and leaving the store is logged, Amazon will be able to use this data to make recommendations for your next visit in a way similar to their online recommendation system. However, there may well be issues about how much we value our privacy, especially given aspects such as the possibility mentioned in the patent application of using facial recognition systems to identify customers.

Netflix

Another Silicon Valley company, Netflix, started in 1997 as a postal DVD rental company. You took out a DVD and added another to your queue, and they would then be sent out in turn. Rather usefully, you had the ability to prioritize your queue. This service is still available and still lucrative, though it appears to be gradually winding down. Now an international, Internet streaming, media provider with approximately seventy-five million subscribers across 190 different countries, in 2015 Netflix successfully expanded into providing its own original programmes.

Netflix collects and uses huge amounts of data to improve customer service, such as offering recommendations to individual customers while endeavouring to provide reliable streaming of its movies. Recommendation is at the heart of the Netflix business model and most of its business is driven by the data-based recommendations it is able to offer customers. Netflix now tracks what you watch, what you browse, what you search for, and the day and time you do all these things. It also records whether you are using an iPad, TV, or something else.

In 2006, Netflix announced a crowdsourcing competition aimed at improving their recommender systems. They were offering a $1 million prize for a collaborative filtering algorithm that would improve by 10 per cent the prediction accuracy of user movie ratings. Netflix provided the training data, over 100 million items, for this machine learning and data mining competition—and no other sources could be used. Netflix offered an interim prize (the Progress Prize) worth $50,000, which was won by the Korbell team in 2007 for solving a related but somewhat easier problem. Easier is a relative term here, since their solution combined 107 different algorithms to come up with two final algorithms, which, with ongoing development, are still being used by Netflix. These algorithms were gauged to cope with 100 million ratings as opposed to the five billion that the full prize algorithm would have had to be able to manage. The full prize was eventually awarded in 2009 to the BellKor's Pragmatic Chaos team whose algorithm represented a 10.06 per cent improvement over the existing one. Netflix never fully implemented the winning algorithm, primarily because by this time their business model had changed to the now-familiar one of media streaming.

Once Netflix expanded their business model from a postal service to providing movies by streaming, they were able to gather a lot more information on their customers' preferences and viewing habits, which in turn enabled them to provide improved recommendations. However, in a departure from the digital modality, Netflix employs part-time *taggers*, a total of about forty people worldwide who watch movies and tag the content, labelling them as, for example, 'science fiction' or 'comedy'. This is how films get categorized—using human judgement and not a computer algorithm initially; that comes later.

Netflix uses a wide range of recommender algorithms that together make up a recommender system. All these algorithms act on the aggregated big data collected by the company. Content-based filtering, for example, analyses the data reported by the 'taggers' and finds similar movies and TV programmes according to criteria such as genre and actor. Collaborative filtering monitors such things as your viewing and search habits. Recommendations are based on what viewers with similar profiles watched. This was less successful when a user account has more than one user, typically several members of a family, with inevitably different tastes and viewing habits. In order to overcome this problem, Netflix created the option of multiple profiles within each account.

On-demand Internet TV is another area of growth for Netflix and the use of big data analytics will become increasingly important as they continue to develop their activities. As well as collecting search data and star ratings, Netflix can now keep records on how often users pause or fast forward, and whether or not they finish watching each programme they start. They also monitor how, when, and where they watched the programme, and a host of other variables too numerous to mention. Using big data analytics we are told that they are now even able to predict quite accurately whether a customer will cancel their subscription.

Data science

'Data scientist' is the generic title given to those who work in the field of big data. The McKinsey Report of 2012 highlighted the lack of data scientists in the USA alone, estimating that by 2018 the shortage would reach 190,000. The trend is apparent worldwide and even with government initiatives promoting data science skills training, the gap between available and required expertise seems to be widening. Data science is becoming a popular study option in universities but graduates so far have been unable to meet the demands of commerce and industry, where positions in data science offer high salaries to experienced applicants. Big data for commercial enterprises is concerned with profit, and disillusionment will set in quickly if an over-burdened data analyst with insufficient experience fails to deliver the expected positive results. All too often, firms are asking for a one-size-fits-all model of data scientist who is expected to be competent in everything from statistical analysis to data storage and data security.

Data security is of crucial importance to any firm and big data creates its own security issues. In 2016, the Netflix Prize 2 initiative was cancelled because of data security concerns. Other recent data hacks include Adobe in 2013, eBay and JP Morgan Chase Bank in 2014, Anthem (a US health insurance company) and Carphone Warehouse in 2015, MySpace in 2016, and LinkedIn—a 2012 hack not discovered until 2016. This is a small sample; many more companies have been hacked or suffered other types of security breaches leading to the unauthorized dissemination of sensitive data. In Chapter 7, we will look at some of the big data security breaches in depth.

Chapter 7
Big data security and the Snowden case

In July 2009, Amazon Kindle readers found life imitating art when their copy of Orwell's novel *1984* completely disappeared from their devices. In *1984*, the 'memory hole' is used to incinerate documents that are considered subversive or no longer wanted. Documents permanently disappear and history is rewritten. It could almost have been an unfortunate prank but *1984* and Orwell's *Animal Farm* had actually been removed as the result of a dispute between Amazon and the publisher. Customers were angry, having paid for the e-book and assumed that it was therefore their property. A lawsuit filed by a highschool student and one other person was settled out of court. In the settlement, Amazon stated that they would no longer erase books from people's Kindles, except in certain circumstances, including that '*a judicial or regulatory order requires such deletion or modification*'. Amazon offered customers a refund, gift certificate, or to restore the deleted books. In addition to being unable to sell or to lend our Kindle books, it seems we do not actually own them at all.

Although the Kindle incident was in response to a legal problem and was not intended maliciously, it serves to illustrate how straightforward it is to delete e-documents, and without hard copies, how simple it would be to completely eradicate any text viewed as undesirable or subversive. If you pick up the physical version of this book tomorrow and read it you know with absolute certainty it will be the same as it was today but if you read anything on the Web today, you cannot be certain that it will be the same when you read it tomorrow. There is no absolute certainty on the Web. Since e-documents can be modified and updated without the author's knowledge, they can easily be manipulated. This situation could be extremely damaging in many different situations, such as the possibility of someone tampering with electronic medical records. Even digital signatures, designed to authenticate electronic documents, can be hacked. This highlights some of the problems facing big data systems, such as ensuring they actually work as intended, can be fixed when they break down, and are tamper-proof and accessible only to those with the correct authorization.

Securing a network and the data it holds are the key issues here. A basic measure taken to safeguard networks against unauthorized access is to install a *firewall*, which isolates a network from unauthorized outside access through the Internet. Even if a network is secure from direct attack, for example from viruses and trojans, the data stored on it, particularly if it is unencrypted, can still be compromised. For instance, one such technique, that of phishing, attempts to introduce malicious code, usually by sending an email with an executable file or requesting personal or security data such as passwords. But the main problem facing big data is that of hacking.

The retail store Target was hacked in 2013 leading to the theft of the details of an estimated 110 million customer records, including credit card details of forty million people. It is reported that by the end of November the intruders had successfully pushed their malware to most of Target's point-of-sale machines and were able to collect customer card records from real-time transactions. At that time, Target's security system was being monitored twenty-four hours a day by a team of specialists working in Bangalore. Suspicious activity was flagged and the team contacted the primary security team located in Minneapolis, who unfortunately failed to act on the information. The Home Depot hack, which we will look at next, was even bigger but used similar techniques, leading to a massive data theft.

Home Depot hack

On 8 September 2014, Home Depot, which describes itself as the largest home improvement retailer in the world, announced in a press release that its payment data systems had been hacked. In an update on 18 September 2014, Home Depot reported that the attack had affected approximately fifty-six million debit/credit cards. In other words, fifty-six million debit/credit cards details were stolen. In addition, fifty-three million email addresses were also stolen. In this case, the hackers were able to first steal a vendor's log, giving them easy access to the system—but only to the individual vendor's part of the system. This was accomplished by a successful phishing attempt.

The next step required the hackers to access the extended system. At that time, Home Depot was using Microsoft XP operating system, which contained an inherent flaw that the hackers exploited. The self-checkout system was then targeted since this sub-system was itself clearly identifiable within the entire system. Finally, the hackers infected the 7,500 self-checkout terminals with malware to gain customer information. They used BlackPOS, also known as Kaptoxa, a specific malware for scraping credit/debit card information from infected terminals. For security, payment card information should be encrypted when the card is swiped at a point-of-sales terminal but apparently this feature, point-to-point encryption, had not been implemented and so the details were left open for the hackers to take.

The theft was uncovered when banks started to detect fraudulent activity on accounts that had made other recent purchases at Home Depot—the card details had been sold through Rescator, a cybercrime outlet found on the dark Web. It is interesting that people using cash registers, which also take cards, were not affected by this attack. The reason for this appears to be that in the mainframe computer, cash registers were identified only by numbering and so were not readily identifiable as checkout points by the criminals. If Home Depot had also used simple numbering for its self-checkout terminals, this hacking attempt might have been foiled. Having said that, at the time Kaptoxa was deemed state-of-the-art malware and was virtually undetectable, so given the open access to the system the hackers had obtained, it almost certainly would eventually have been introduced successfully.

The biggest data hack yet

In December 2016, Yahoo! announced that a data breach involving over one billion user accounts had occurred in August 2013. Dubbed the biggest ever cyber theft of personal data, or at least the biggest ever divulged by any company, thieves apparently used forged cookies, which allowed them access to accounts without the need for passwords. This followed the disclosure of an attack on Yahoo! in 2014, when 500 million accounts were compromised. Chillingly, Yahoo! alleged the 2014 hack was perpetrated by an unnamed 'state-sponsored actor'.

Cloud security

The list of big data security breaches increases almost daily. Data theft, data ransom, and data sabotage are major concerns in a data-centric world. There have been many scares regarding the security and ownership of personal digital data. Before the digital age we used to keep photos in albums and negatives were our backup. After that, we stored our photos electronically on a hard-drive in our computer. This could possibly fail and we were wise to have back-ups but at least the files were not publicly accessible. Many of us now store data in the Cloud. Photos, videos, home movies all require a lot of storage space and so the Cloud makes sense from that perspective. When you store your files in the Cloud, you are uploading them to a data centre—more likely, they will be distributed across several centres—and more than one copy will be kept.

If you store all your photos in the Cloud, it's highly unlikely with today's sophisticated systems that you would lose them. On the other hand, if you want to delete something, maybe a photo or video, it becomes difficult to ensure all copies have been deleted. Essentially you have to rely on your provider to do this. Another important issue is controlling who has access to the photos and other data you have uploaded to the Cloud. If we want to make big data secure, encryption is vital.

Encryption

Encryption, as mentioned briefly in Chapter 5, refers to methods used to scramble files so that they cannot easily be read, and the basic technique goes back at least as far as Roman times. Suetonius, in his *The Twelve Caesars*, describes how Julius Caesar encoded documents using a three-letter shift to the left. Using this method the word 'secret' would be encoded as 'pbzobq'. Known as a 'Caesar cipher' this is not difficult to break, but even the most secure ciphers used today apply shifting as part of the algorithm.

In 1997, the best publicly available encryption method, Data Encryption Standard (DES), was shown to be breakable, largely due to the increase in computing power available and the relatively short 56-bit key length. Although this provides a possible 2^{56} different key choices, it was possible to decrypt a message by testing each one until the correct key was found. This was done in 1998, in just over twenty-two hours using Deep Crack, a computer built by Electronic Frontier Foundation expressly for this purpose.

In 1997, the National Institute of Standards and Technology (NIST) in the USA, concerned that DES lacked the security needed for protecting top secret documents, launched an open, worldwide competition to find a better encryption method than DES. The competition ended in 2001 with the AES algorithm being chosen. It was submitted as the Rijndael algorithm, combining the names of its two Belgian originators, Joan Daemen and Vincent Rijmen.

AES is a software algorithm used for text encryption with a choice of a 128-, 192-, or 256-bit key length. For the 128-bit key length, the algorithm requires nine processing rounds, each consisting of four steps, plus a final round with only three steps. The AES encryption algorithm is iterative and performs a large number of computations on matrices—just the kind of calculations that are best performed by computers. However, we can describe the process informally without reference to the mathematical transformations.

AES starts by applying a key to the text we want to encrypt. We would no longer be able to recognize the text but given the key, we could easily decode it so more steps are needed. The next step involves substituting each letter with another letter, using a special look-up table, called a Rijndael S-Box. Again, if we have the Rijndael S-Box, we can work backwards to decrypt the message. A Caesar Cipher, where letters are shifted to the left, and a final permutation completes one round. The result is then used to start another round, using a different key and so on, until all rounds have been completed. Of course, we have to be able to decrypt, and for this algorithm the method is reversible.

For the 192-bit key length there are twelve rounds in total. For even greater security, a longer key length, AES 256, can be employed, but most users, including Google and Amazon, find AES 128 sufficient for their big data security needs. AES is secure and has yet to be broken, leading to several governments to ask major companies such as Apple and Google to provide back doors into the encrypted material.

Big data security and the Snowden case

95

Email security

It has been estimated that in 2015 over 200 billion emails were sent every day, with less than 10 per cent of these being authentic and not spam or with malicious intent. Most emails are not encrypted, making their contents vulnerable to interception by hackers. When I send an unencrypted email, let's say from California to the UK for example, it is divided into data 'packets' and transmitted through a mail server, which is connected to the Internet. The Internet is essentially made up of a big worldwide network of wires, above ground, below ground, and below oceans, plus cell phone towers and satellites. The only continent unconnected by transoceanic cables is Antarctica.

So although the Internet and Cloud-based computing are generally thought of as wireless, they are anything but; data is transmitted through fibre-optic cables laid under the oceans. Nearly all digital communication between continents is transmitted in this way. My email will be sent via transatlantic fibre-optic cables, even if I am using a Cloud computing service. The Cloud, an attractive buzz word, conjures up images of satellites sending data across the world, but in reality Cloud services are firmly rooted in a distributed network of data centres providing Internet access, largely through cables.

Fibre-optic cables provide the fastest means of data transmission and so are generally preferable to satellites. The current extensive research into fibre-optic technology is resulting in ever faster data transmission rates. Transatlantic cables have been the target of some curious and unexpected attacks, including those from sharks intent on biting through the cables. While, according to the International Cable Protection Committee, shark attacks account for fewer than 1 per cent of the faults logged, even so, cables in vulnerable areas are now often protected using Kevlar. Assuming there are no problems with transatlantic cables due to inquisitive sharks, hostile governments, or careless fishermen, and my email makes landfall in the UK and continues on its way, it may be at this point that, as with other Internet data, it is intercepted. In June 2013, Edward Snowden leaked documents revealing that the Government Communications Headquarters (GCHQ) in the UK were tapping into a vast amount of data, received through approximately 200 transatlantic cables, using a system called Tempora.

The Snowden case

Edward Snowden is an American computer professional who was charged with espionage in 2013 after leaking classified information from the US National Security Agency (NSA). This high-profile case brought government mass surveillance capabilities to the attention of the general public, and widespread concerns were expressed regarding individual privacy. Awards made to Snowden since taking this action have been many and include election as rector of the University of Glasgow, the *Guardian*'s Person of the Year 2013, and Nobel Peace Prize nominations in 2014, 2015, and 2016. He has the support of Amnesty International as a whistleblower who provided a service to his country. However, US government officials and politicians have begged to differ in this view.

In June 2013, the *Guardian* newspaper in the UK reported that the NSA was collecting metadata from some of the major US phone networks. This report was swiftly followed by the revelation that a program called PRISM was being used to collect and store Internet data on foreign nationals communicating with the US. A whole slew of electronic leaks followed, incriminating both the US and the UK. A Booz Allen Hamilton employee and NSA contractor working at the Hawaii Cryptologic Center, Edward Snowden, was the source of these leaks, which he sent to members of the media he felt could be trusted not to publish without careful consideration. Snowden's motivations, and the legal issues involved, are beyond the scope of this book but it is apparent that he believed that what had started out as legitimate spying on other countries had now turned in on itself and the NSA was now spying, illegally, on all US citizens.

The free Web scraping tools, DownThemAll, an available extension of Mozilla Firefox, and the program *wget*, give the means to quickly download the entire contents of a website or other Web-related data. These applications, available to authorized users on NSA classified networks, were used by Snowden to download and copy massive amounts of information. He also transferred large amounts of highly sensitive data from one computer system to another. In order to do this, he needed usernames and passwords, which a systems administrator would routinely have. He thus had easy access to many of the classified documents he stole, but not all. To get access to higher than top-secret documents, he had to use the authentication details of higher level user accounts, which security protocols should have prevented. However, since he had created these accounts and had system administrator privileges, he knew the account details. Snowden also managed to persuade at least one NSA employee with security clearance higher than his to tell him their password.

Ultimately, Snowden copied an estimated 1.5 million highly classified documents, of which about 200,000 (Snowden understood that not all of his stolen documents should be made public and was cautious about which should be published) were handed over to trusted reporters, although relatively few of even these were eventually published.

While the details have never been fully revealed by Snowden, it seems he was able to copy the data onto flash drives, which he apparently had no difficulty in taking with him when he left work for the day. Security measures to prevent Snowden from being able to remove these documents were clearly inadequate. Even a simple body scan on exiting the facility would have detected any portable devices, and video surveillance in the offices could also have flagged suspicious activity. In December 2016, the US House of Representatives declassified a document dated September 2016, which remains heavily redacted, reviewing Snowden the man as well as the nature and impact of the leaked documents. From this document it is clear that the NSA had not applied sufficient security measures and as a result the Secure the Net initiative has since been put into operation, although it is yet to be fully implemented.

Snowden had extensive system administrator privileges, but given the extremely sensitive nature of the data, allowing one person to have full access with no safeguards was not acceptable. For example, requiring validation credentials of two people when data was accessed or transferred might have been sufficient to prevent Snowden from illicitly copying files. It is also curious that Snowden could apparently plug in a USB drive and copy anything he wanted. A very simple security measure is to disable DVD and USB ports or not have them installed in the first place. Add further authentication using retina scan to the requirement for a password and it would have been very difficult for Snowden even to access those higher level documents. Modern security techniques are sophisticated and difficult to penetrate if used correctly.

In late 2016, entering 'Edward Snowden' in Google search gave over twenty-seven million results in just over one second; and the search term 'Snowden' gave forty-five million results. Since many of these sites give access to or even display the leaked documents labelled 'Top Secret', they are now firmly in the global public domain and will no doubt remain so. Edward Snowden is currently living in Russia.

In contrast with Edward Snowden's case, WikiLeaks presents a very different story.

WikiLeaks

WikiLeaks is a huge online whistleblowing organization whose aim is to disseminate secret documents. It is funded by donations and staffed largely by volunteers, though it does appear to employ a few people too. As of December 2015, WikiLeaks claims to have published (or leaked) more than ten million documents. WikiLeaks maintains its highly public profile through its own website, Twitter, and Facebook.

Highly controversial, WikiLeaks and its leader Julian Assange hit the headlines on 22 October 2010 when a vast amount of classified data—391,832 documents—dubbed 'Iraq War Logs' was made public. This followed the approximately 75,000 documents constituting 'The Afghan War Diary' that had already been leaked on 25 July 2010.

An American army soldier, Bradley Manning, was responsible for both leaks. Working as an intelligence analyst in Iraq, he took a compact disc to work with him and copied secret documents from a supposedly secure personal computer. For this, Bradley Manning, now known as Chelsea Manning, was sentenced in 2013 to thirty-five years in prison following conviction, by court-martial, for violations of the Espionage Act and other related offences. Former US president Barack Obama commuted Chelsea Manning's sentence in January 2017, prior to his leaving office. Ms Manning, who received treatment for gender dysphoria while in prison, was released on 17 May 2017.

Heavily criticized by politicians and governments, WikiLeaks has nonetheless been applauded by and received awards from the likes of Amnesty International (2009) and the UK's *The Economist* (2008), among a long list of others. According to their website, Julian Assange has been nominated for the Nobel Peace Prize in six consecutive years, 2010–15. The Nobel

Committee does not release the names of nominees until fifty years have passed but nominators, who have to meet the strict criteria of the Peace Prize committee, often do publicly announce the names of their nominees. For example, in 2011, Julian Assange was nominated by Snorre Valen, a Norwegian parliamentarian, in support of WikiLeaks exposing alleged human rights violations. In 2015, Assange had the support of former UK member of parliament George Galloway, and in early 2016 a supportive group of academics also called for Assange to be awarded the prize.

Yet by the end of 2016, the tide was turning against Assange and WikiLeaks, at least in part because of alleged bias in their reporting. Against WikiLeaks are ethical concerns regarding the safety and privacy of individuals; corporate privacy; government secrecy; the protection of local sources in areas of conflict; and the public interest in general. The waters are becoming increasingly muddied for Julian Assange and WikiLeaks. For example, in 2016, emails were leaked at a time best suited to damage Hillary Clinton's presidential candidacy, raising questions about WikiLeaks' objectivity, and prompting considerable criticism from a number of well-respected sources.

Regardless of whether you support or condemn the activities of Julian Assange and WikiLeaks, and almost inevitably people will do both, varying with the issue at stake, one of the big technical questions is whether it is possible to shut down WikiLeaks. Since it maintains its data on many servers across the world, some of it in sympathetic countries, it is unlikely that it could be completely shut down, even assuming that this was desirable. However, for increased protection against retaliation following each disclosure, WikiLeaks has issued an insurance file. The unspoken suggestion is that if anything happens to Assange or if WikiLeaks is shut down, the insurance file key will be publicly broadcast. The most recent WikiLeaks insurance file uses AES with a 256-bit key and so it is highly unlikely to be broken.

As of 2016, Edward Snowden is at odds with WikiLeaks. The disagreement comes down to how each of them managed their data leaks. Snowden handed his files over to trusted journalists, who carefully chose which documents to leak. US government officials were informed in advance, and, following their advice, further documents were withheld because of national security concerns. To this day, many have never been disclosed. WikiLeaks appears simply to publish its data with little effort to protect personal information. It still seeks to gather information from whistleblowers, but it is not clear how reliable recent data leaks have been, or indeed whether its selection of the information it presents allows it to be completely disinterested. On its website, WikiLeaks gives instruction for how to use a facility called TOR (The Onion Router) to send data anonymously and ensure privacy, but you do not have to be a whistleblower to use TOR.

TOR and the dark Web

Janet Vertesi, an assistant professor in the Sociology Department at Princeton University, decided to conduct a personal experiment to see if she could keep her pregnancy a secret from online marketers and so prevent her personal information becoming part of the big data pool. In an article published in *TIME* magazine in May 2014, Dr Vertesi gives an account of her experience. She took exceptional privacy measures, including avoiding social media; she downloaded TOR and used it to order many baby-related items; and in-store purchases were paid for in cash. Everything she did was perfectly legal but ultimately she concluded that opting out was costly and time-consuming and made her look, in her own words, like a 'bad citizen'. However, TOR is worth looking at, not least because it made Dr Vertesi feel safe and maintained her privacy from trackers.

TOR is an encrypted network of servers that was originally developed by the US Navy to provide a way of using the

Internet anonymously, and so prevent tracking and the collection of personal data. TOR is an ongoing project, aimed at developing and improving open-source online anonymity environments that anyone concerned about privacy can use. TOR works by encrypting your data, including the sending address, and then anonymizes it by removing part of the header, crucially including the IP address, since an individual can easily be found by back-tracking given that information. The resulting data package is routed through a system of servers or relays, hosted by volunteers, before arriving at its final destination.

On the positive side, users include the military who originally designed it; investigative journalists wishing to protect their sources and information; and everyday citizens wishing to protect their privacy. Businesses use TOR to keep secrets from other businesses; and governments use it to protect sources of sensitive information as well as the information itself. A TOR Project press release gives a list of some of the news items involving TOR between 1999 and 2016.

On the negative side, the TOR anonymity network has been widely used by cyber criminals. Websites are accessible through TOR-hidden services and have the suffix '.onion'. Many of these are extremely unpleasant, including illegal dark websites used for drug dealing, pornography, and money laundering. For example, the highly publicized website Silk Road, part of the dark Web and a supplier of illegal drugs, was accessed through TOR, making it difficult for law enforcement to track it. A major court case followed the arrest of Ross William Ulbricht, who was subsequently convicted of creating and running Silk Road, using the pseudonym Dread Pirate Roberts. The website was closed down but later sprang back up again, and in 2016 was in its third reincarnation under the name Silk Road 3.0.

Deep Web

The deep Web refers to all those websites that cannot be indexed by the usual search engines, such as Google, Bing, and Yahoo! It comprises legitimate sites as well as those that make up the dark Web. It is popularly estimated to be vastly bigger than the familiar surface Web, though even with special deep Web search engines it is difficult to estimate the size of this hidden world of big data.

Chapter 8
Big data and society

Robots and jobs

The eminent economist, John Maynard Keynes, writing during the British economic depression in 1930, speculated on what working life would be like a century later. The industrial revolution had created new city-based jobs in factories and transformed what had been a largely agrarian society. It was thought that labour-intensive work would eventually be performed by machines, leading to unemployment for some and a much-reduced working week for others. Keynes was particularly concerned with how people would use their increased leisure time, freed from the exigencies of gainful employment by technological advances. Perhaps more pressing was the question of financial support leading to the suggestion that a universal basic income would provide a way of coping with the decline in available jobs.

Gradually over the 20th century we have seen jobs in industry eroded by ever-more sophisticated machines, and although, for example, many production lines were automated decades ago, the Keynesian fifteen-hour working week has yet to materialize and seems unlikely to do so in the near future. The digital revolution will inevitably change employment, just as the industrial revolution did, but in ways we are unlikely to be able to predict accurately. As the technology of the 'Internet of Things' advances, our world continues to become more data-driven. Using the results of real-time big data analysis to inform decisions and actions will play an increasingly important role in our society.

There are suggestions that people will be needed to build and code machines, but this is speculative and, in any case, is just one area of specialized work where we can realistically expect to see robots increasingly taking the place of people. For example, sophisticated robotic medical diagnosis would reduce the medical workforce. Robotic surgeons, with extended Watson-like capabilities, are likely. Natural language processing, another big data area, will develop to the point where we cannot tell whether we are talking to a robotic device or a doctor—at least, when we are not face-to-face.

However, predicting what jobs humans will be doing once robots have taken over many of the existing roles is difficult. Creativity is supposedly the realm of humans, but computer scientists, working in collaboration at the Universities of Cambridge and Aberystwyth, have developed Adam, a robot scientist. Adam has successfully formulated and tested new hypotheses in the field of genomics, leading to new scientific discoveries. The research has progressed with a team at the University of Manchester successfully developing Eve, a robot that works on drug design for tropical diseases. Both these projects implemented artificial intelligence techniques.

The craft of the novelist appears to be uniquely human, relying on experience, emotion, and imagination, but even this area of creativity is being challenged by robots. The Nikkei Hoshi Shinichi Literary Award accepts novels written or co-written by non-human authors. In 2016, four novels written jointly by people and computers passed the first stage of the competition, without the judges knowing the details regarding authorship.

Although scientists and novelists may eventually work collaboratively with robots, for most of us the impact of our big data driven environment will be more apparent in our daily activities, through smart devices.

Smart vehicles

On 7 December 2016, Amazon announced that it had made its first commercial drone delivery using GPS (global positioning system) to find its way. The recipient, a man living in the countryside near Cambridge in the UK, received a package weighing 4.7 pounds. Drone deliveries can currently be made to only two Amazon Prime Air customers, both living within 5.2 square miles of the fulfilment centre near Cambridge. A video, referenced in the Further reading section, shows the flight. This seems likely to signal the start of big data collection for this program.

Amazon is not the first to make a successful commercial drone delivery. In November 2016, Flirtey Inc. started a drone delivery pizza service in a small area from their home base in New Zealand and there have been similar projects elsewhere. At present, it seems likely that drone delivery services will grow, particularly in remote areas where it might be possible to manage safety issues. Of course, a cyber-attack or simply a breakdown in the computer systems could create havoc: if, for example, a small delivery drone were to malfunction, it could cause injury or death to humans or animals, as well as considerable damage to property.

This is what happened when the software controlling a car travelling along the road at 70 mph was taken over remotely. In 2015, two security experts, Charlie Miller and Chris Valasek, working for *Wired* magazine, demonstrated on a willing victim that Uconnect, a dashboard computer used to connect a vehicle to the Internet, could be hacked remotely while the vehicle was in motion. The report makes alarming reading; the two expert hackers were able to use a laptop Internet connection to control the steering, brakes, and transmission along with other less critical functions such as the air-conditioning and radio of a Jeep Cherokee. The Jeep was travelling at 70 mph on a busy public road when suddenly all response to the accelerator failed, causing considerable alarm to the driver.

As a result of this test, the car manufacturer Chrysler issued a warning to the owners of 1.4 million vehicles and sent out USB drives containing software updates to be installed through a port on the dashboard. The attack was made because of a vulnerability in the smartphone network that was subsequently fixed, but the story serves to illustrate the point that the potential for cyber-attacks on smart vehicles will need to be addressed before the technology becomes fully public.

The advent of autonomous vehicles, from cars to planes, seems inevitable. Planes can already fly themselves, including taking off and landing. Although it's a step away to think of drones being in widespread use for transporting human passengers, they are currently used in farming for intelligent crop spraying and also for military purposes. Smart vehicles are still in the early stages of development for general use but smart devices are already part of the modern home.

Smart homes

As mentioned in Chapter 3, the term 'Internet of Things' (IoT) is a convenient way of referring to the vast numbers of electronic sensors connected to the Internet. For example, any electronic device that can be installed in a home and managed remotely—through a user interface displayed on the resident's television screen, smartphone, or laptop—is a smart device and so part of the IoT. Voice-activated central control points are installed in many homes that manage lighting, heating, garage doors, and many other household devices. Wi-Fi (which stands for 'wireless fidelity', or the capacity to connect with a network, like the Internet, using radio waves rather than wires) connectivity means that you can ask your smart speaker (by its name, which you will have given it) for the local weather or national news reports.

These devices provide Cloud-based services, and are not without their drawbacks when it comes to privacy. As long as the device is switched on, everything you say is recorded and stored in a remote server. In a recent murder investigation, police in the United States asked Amazon to release data from an Echo device (which is voice controlled and connects to the Alexa Voice Service to play music, provide information, news reports, etc.) that they believed would assist them in their inquiries. Amazon was initially unwilling to do so, but the suspect has recently given his permission for them to release the recordings, hoping that they will help prove his innocence.

Further developments, based on Cloud computing, mean that electrical appliances such as washing machines, refrigerators, and home-cleaning robots will be part of the smart home and managed remotely through smartphones, laptops, or home speakers. Since all these systems are Internet controlled they are potentially at risk from hackers, and so security is a big area of research.

Even children's toys are not immune. Named '2014 Innovative Toy of the Year' by the London Toy Industry Association, a smart doll called 'My Friend Cayla' was subsequently hacked. Through an unsecured bluetooth device hidden in the doll, a child can ask the doll questions and hear replies. The German Federal Network Agency, responsible for monitoring Internet communications, has encouraged parents to destroy the doll, which has now been banned, because of the threat to privacy that it presents. Hackers have been able to show that it is fairly easy listen to a child and provide inappropriate answers, including words from the manufacturer's banned list.

Smart cities

Although the smart home is only just becoming a reality, the IoT together with multiple information and communication technologies (ICTs) are now predicted to make smart cities a reality. Many countries, including India, Ireland, the UK, South Korea, China, and Singapore, are already planning smart cities. The idea is that of greater efficiency in a crowded world since cities are growing rapidly. The rural population is moving to the city at an ever-increasing rate. In 2014, about 54 per cent lived in cities and by 2050 the United Nations predicts that about 66 per cent of the world's population will be city dwellers.

The technology of smart cities is propelled by the separate but accumulating ideas from early implementations of the IoT and big data management techniques. For example, driverless cars, remote health monitoring, the smart home, and tele-commuting would all be features of a smart city. Such a city would depend on the management and analysis of the big data accumulated from the sum total of the city's vast sensor array. Big data and the IoT working together are the key to smart cities.

For the community as a whole, one of the benefits would be a smart energy system. This would regulate street lighting, monitor traffic, and even track garbage. All this could be achieved by installing a huge array of radio-frequency identification (RFID) tags and wireless sensors across the city. These tags, which consist of a microchip and a tiny antenna, would send data from individual devices to a central location for analysis. For example, the city governance would monitor traffic by installing RFID tags on vehicles and digital cameras on streets. Improved personal safety would also be a consideration, for example, children could be discretely tagged and monitored through their parents' cell phones. These sensors would create a huge amount of data which would need to be monitored and analysed in real-time, through a central data processing unit. It could then be used for a variety of purposes including gauging traffic flow, identifying congestion, and recommending alternative routes. Data security would clearly be of paramount importance in this context, as any major breakdown in the system or hacking would quickly affect public confidence.

Songdo International Business District in South Korea, scheduled for completion in 2020, has been purpose built as a smart city. One of the main features is that the entire city has fibre-optic broadband. This state-of-the-art technology is used to ensure the desired features of a smart city can be accessed quickly. New smart cities are also being designed to minimize negative environmental effects, making them the sustainable cities of the future. While many smart cities have been planned and, like Songdo, are being purpose built, existing cities will need to modernize their infrastructures gradually.

In May 2016, the United Nations Global Pulse, an initiative aimed at promoting big data research for global benefit, unveiled its open 'Big Ideas Competition 2016: Sustainable Cities' for the ten member states of the Association of Southeast Asian Nations (ASEAN) and the Republic of Korea. By the June deadline, over 250 proposals had been received and the winners in various categories were announced in August 2016. The Grand Prize went to the Republic of Korea for their proposal to improve public transport by reducing waiting times by utilizing crowd-sourcing information on queues.

Looking forward

In this *Very Short Introduction*, we have seen how the science of data has undergone a radical transformation over the past few decades due to the technological advances brought about by the development of the Internet and the digital universe. In this final chapter, we have glimpsed some of the ways our lives may be shaped by big data in the future. While we can't hope to cover in a short introduction all the areas in which big data is making an impact, we have seen some of the diverse applications that already affect us.

The data generated by the world is only going to get bigger. Methods for dealing with all this data effectively and meaningfully will undoubtedly continue to be the subject of intense research, particularly in the area of real-time analysis. The big data revolution marks a sea-change in the way the world works, and as with all technological developments, individuals, scientists, and governments together have a moral responsibility to ensure its proper use. Big data is power. Its potential for good is enormous. How we prevent its abuse is up to us.

Big Data

Byte size chart

Term	Meaning
Bit	1 binary digit: 0 or 1
Byte	8 bits
Kilobyte (Kb)	1,000 bytes
Megabyte (Mb)	1,000 kilobytes
Gigabyte (Gb)	1,000 megabytes
Terabyte (Tb)	1,000 gigabytes
Petabyte (Pb)	1,000 terabytes
Exabyte (Eb)	1,000 petabytes
Zettabyte (Zb)	1,000 exabytes
Yottabyte (Yb)	1,000 zettabytes

ASCII table for lower case letters

Decimal	Binary	Hex	Letter
97	01100001	61	a
98	01100010	62	b
99	01100011	63	c
100	01100100	64	d
101	01100101	65	e
102	01100110	66	f
103	01100111	67	g
104	01101000	68	h
105	01101001	69	i
106	01101010	6A	j
107	01101011	6B	k
108	01101100	6C	l
109	01101101	6D	m
110	01101110	6E	n
111	01101111	6F	o

Decimal	Binary	Hex	Letter
112	01110000	70	p
113	01110001	71	q
114	01110010	72	r
115	01110011	73	s
116	01110100	74	t
117	01110101	75	u
118	01110110	76	v
119	01110111	77	w
120	01111000	78	x
121	01111001	79	y
122	01111010	7A	z
32	00010000	20	space

Further reading

Chapter 1: The data explosion

David J. Hand, *Information Generation: How Data Rule Our World* (Oneworld, 2007).

Jeffrey Quilter and Gary Urton (eds), *Narrative Threads: Accounting and Recounting in Andean Khipu* (University of Texas Press, 2002).

David Salsburg, *The Lady Tasting Tea: How Statistics Revolutionized Science in the Twentieth Century* (W.H. Freeman and Company, 2001).

Thucydides, *History of the Peloponnesian War*, ed. and intro. M. I. Finley, trans. Rex Warner (Penguin Classics, 1954).

Chapter 2: Why is big data special?

Joan Fisher Box, *R. A. Fisher: The Life of a Scientist* (Wiley, 1978).

David J. Hand, *Statistics: A Very Short Introduction* (Oxford University Press, 2008).

Viktor Mayer-Schönberger and Kenneth Cukier, *Big Data: A Revolution That Will Transform How We Live, Work, and Think* (Mariner Books, 2014).

Chapter 3: Storing big data

C. J. Date, *An Introduction to Database Systems* (8th edn; Pearson, 2003).

Guy Harrison, *Next Generation Databases: NoSQL and Big Data* (Springer, 2015).

Chapter 4: Big data analytics

Thomas S. Kuhn and Ian Hacking, *The Structure of Scientific Revolutions: 50th Anniversary Edition* (University of Chicago Press, 2012).

Bernard Marr, *Big Data: Using SMART Big Data, Analytics and Metrics to Make Better Decisions and Improve Performance* (Wiley, 2015).

Lars Nielson and Noreen Burlingame, *A Simple Introduction to Data Science* (New Street Communications, 2012).

Chapter 5: Big data and medicine

Dorothy H. Crawford, *Ebola: Profile of a Killer Virus* (Oxford University Press, 2016).

N. Generous, G. Fairchild, A. Deshpande, S. Y. Del Valle, and R. Priedhorsky, 'Global Disease Monitoring and Forecasting with Wikipedia', *PLoS Comput Biol* 10(11) (2014), e1003892. doi: 10.1371/journal.pcbi.1003892

Peter K. Ghavami, 'Clinical Intelligence: The Big Data Analytics Revolution in Healthcare. A Framework for Clinical and Business Intelligence' (PhD thesis, 2014).

D. Lazer and R. Kennedy, 'The Parable of Google Flu: Traps in Big Data Analysis', *Science* 343 (2014), 1203–5. <http://scholar.harvard.edu/files/gking/files/0314policyforumff.pdf>.

Katherine Marconi and Harold Lehmann (eds), *Big Data and Health Analytics* (CRC Press, 2014).

Robin Wilson, Elizabeth zu Erbach-Schoenberg, Maximilian Albert, Daniel Power et al., 'Rapid and Near Real-Time Assessments of Population Displacement Using Mobile Phone Data Following Disasters: The 2015 Nepal Earthquake', *PLOS Currents Disasters*, Edition 1, 24 Feb 2016, Research Article. doi: 10.1371/currents.dis.d073fbece328e4c39087bc086d694b5c <http://currents.plos.org/disasters/article/rapid-and-near-real-time-assessments-of-population-displacement-using-mobile-phone-data-following-disasters-the-2015-nepal-earthquake/>.

Chapter 6: Big data, big business

Leo Computers Society, *LEO Remembered, By the People Who Worked on the World's First Business Computers* (Leo Computers Society, 2016).

James Marcus, *Amazonia* (The New Press, 2004).

Bernard Marr, *Big Data in Practice* (Wiley, 2016).

Frank Pasquale, *The Black Box Society: The Secret Algorithms That Control Money and Information* (Harvard University Press, 2015).

Foster Provost and Tom Fawcett, *Data Science for Business* (O'Reilly, 2013).

Chapter 7: Big data security and the Snowden case

Andy Greenberg, *This Machine Kills Secrets* (PLUME, 2013).

Glenn Greenwald, *No Place to Hide: Edward Snowden, the NSA, and the U.S. Surveillance State* (Metropolitan Books, 2014).

Luke Harding, *The Snowden Files* (Vintage Books, 2014).

G. Linden, B. Smith, and J. York, 'Amazon.com Recommendations: Item-to-item Collaborative Filtering', *Internet Computing* 7(1) (2003), 76–80.

Fred Piper and Sean Murphy, *Cryptography: A Very Short Introduction* (Oxford University Press, 2002).

P. W. Singer and Allan Friedman, *Cybersecurity and Cyberwar: What Everyone Needs to Know* (Oxford University Press, 2014).

Nicole Starosielski, *The Undersea Network* (Duke University Press, 2015).

Janet Vertesi, 'How Evasion Matters: Implications from Surfacing Data Tracking Online', *Interface: A Special Topics Journal* 1(1) (2015), Article 13. http://dx.doi.org/10.7710/2373-4914.1013 <http://commons.pacificu.edu/cgi/viewcontent.cgi?article=1013& context=interface>.

Chapter 8: Big data and society

Anno Bunnik and Anthony Cawley, *Big Data Challenges: Society, Security, Innovation and Ethics* (Palgrave Macmillan, 2016).

Samuel Greengard, *The Internet of Things* (MIT Press, 2015).

Robin Hanson, *The Age of Em* (Oxford University Press, 2016).

Websites

<https://www.infoq.com/articles/cap-twelve-years-later-how-the-rules-have-changed>

<https://www.emc.com/collateral/analyst-reports/idc-the-digital-universe-in-2020.pdf>

<http://newsroom.ucla.edu/releases/ucla-research-team-invents-new-249693>

<http://www.ascii-code.com/>

<http://www.tylervigen.com/spurious-correlations>

<https://www.statista.com/topics/846/amazon/>

<https://www.wired.com/2015/07/jeep-hack-chrysler-recalls-1-4m-vehicles-bug-fix/>

<http://www.unglobalpulse.org/about-new>

<https://intelligence.house.gov/news/>

<http://www.unglobalpulse.org/about-new>

"牛津通识读本"已出书目

古典哲学的趣味	福柯	地球
人生的意义	缤纷的语言学	记忆
文学理论入门	达达和超现实主义	法律
大众经济学	佛学概论	中国文学
历史之源	维特根斯坦与哲学	托克维尔
设计，无处不在	科学哲学	休谟
生活中的心理学	印度哲学祛魅	分子
政治的历史与边界	克尔凯郭尔	法国大革命
哲学的思与惑	科学革命	丝绸之路
资本主义	广告	民族主义
美国总统制	数学	科幻作品
海德格尔	叔本华	罗素
我们时代的伦理学	笛卡尔	美国政党与选举
卡夫卡是谁	基督教神学	美国最高法院
考古学的过去与未来	犹太人与犹太教	纪录片
天文学简史	现代日本	大萧条与罗斯福新政
社会学的意识	罗兰·巴特	领导力
康德	马基雅维里	无神论
尼采	全球经济史	罗马共和国
亚里士多德的世界	进化	美国国会
西方艺术新论	性存在	民主
全球化面面观	量子理论	英格兰文学
简明逻辑学	牛顿新传	现代主义
法哲学：价值与事实	国际移民	网络
政治哲学与幸福根基	哈贝马斯	自闭症
选择理论	医学伦理	德里达
后殖民主义与世界格局	黑格尔	浪漫主义

批判理论   德国文学   儿童心理学

电影   戏剧   时装

俄罗斯文学   腐败   现代拉丁美洲文学

古典文学   医事法   卢梭

大数据   癌症